高等学校工程管理专业规划教材

国际工程合同管理（双语）

Contract Management for International Construction

韦　嘉　主　编
刘　艳　副主编

中国建筑工业出版社

图书在版编目（CIP）数据

国际工程合同管理（双语）/韦嘉主编. —北京：中国建筑工业出版社，2010
高等学校工程管理专业规划教材
ISBN 978-7-112-11716-1

Ⅰ. 国… Ⅱ. 韦… Ⅲ. 对外承包-承包工程-经济合同-管理-双语教学-高等学校-教材 Ⅳ. F752.68

中国版本图书馆CIP数据核字（2010）第000809号

责任编辑：牛 松 王 跃
责任设计：赵明霞
责任校对：赵 颖

高等学校工程管理专业规划教材
国际工程合同管理（双语）
Contract Management for International Construction
韦 嘉 主 编
刘 艳 副主编

*

中国建筑工业出版社出版、发行（北京西郊百万庄）
各地新华书店、建筑书店经销
北京密云红光制版公司制版
北京建筑工业印刷厂印刷

*

开本：787×1092毫米 1/16 印张：13¾ 字数：402千字
2010年2月第一版 2018年9月第三次印刷
定价：**26.00**元

ISBN 978-7-112-11716-1
(18962)

前　　言

随着我国建筑企业“走出去”的步伐日益加快，我国对国际建筑市场通用的合同管理知识的需求也愈加迫切。近年来，我们从高校毕业生就业导向的调研中也发现：目前愈来愈多的中国涉外建筑企业迫切需要工程管理专业的高校毕业生到国外去从事合同管理的工作，因此，工程合同管理尤其是国际工程合同管理作为一种知识体系对将来从事涉外建筑工程管理的高校毕业生或研究生应该是一门日益重要的必修课。本书不仅适用于工程管理和工程造价专业的本科生及研究生，而且对从事国际工程承包管理的合同管理人员也具有一定的参考价值。

为适应双语教学的要求，同时考虑到该课程的性质比较适合双语教学，因此本书以英文编写。同时，为了更好地帮助读者更深入地理解本书内容，书中对重要的概念、术语以及较难理解的句子进行了详细解释，并有译文。

本书介绍了国际建筑市场常用的合同管理的主要内容，其中包括建筑业的概况；建设项目的参与方；建筑工程的采购方式；招标及合同构成；合同条款；承包商的合同义务；雇主的合同义务；时间；支付；合同管理者的角色；分包；合同的暂停和终止；合同纠纷的解决方法；仲裁与诉讼。

本书在编写过程中得到了中国建筑工业出版社牛松编辑的大力支持和帮助。同时，东南大学李启明教授也在百忙中为本书提出了宝贵的意见，谨在此表示衷心的感谢。

本书由青岛理工大学管理学院副教授韦嘉主编，刘艳任副主编。

由于作者水平有限，书中不当之处在所难免，敬请各位读者和同仁指正。

编　者

2009年12月

Contents

Chapter 1 Introduction to Construction Industry

"To form by assembling parts" is the dictionary definition for construction, but the phrase also is a metaphor for the construction process itself. Just as different materials come together to form a structure, so, too, does a diverse group of people come together to make the project possible. To bring together numerous independent businesses and corporate personalities into one goal oriented process is the peculiar challenge of the construction industry. The organizational cultures of architects, engineers, owners, builders, manufacturers, and suppliers may seem to work against the real need to forge a partnership that will ensure the success of a project. Yet, despite these very real challenges in the industry, construction projects do get completed. In fact, it is difficult to think of an industry that is more basic to our economy and to our daily lives. The highways we drive on, the bridges we cross, the water we drink, the fuel we burn: all are made possible by the activities of the construction industry. Likewise, where we shop, where we work, where we worship and learn, and where we live all exist because of the industry. Designers have visions; but until the contractor builds, those visions are just dreams on a sheet of paper.

Construction is also very intertwined with other aspects of our lives. It affects and is affected by developments in technology, computers, government policies, labor relations, and economic and political practices. Take, for instance, the technological leap of the skyscraper. Until the late 1800s, most buildings were four or five stories high. Masonry supported the structure from the ground. As the height of the building increased, the massing at the ground level also increased to support the additional load. Thus, if the building were built too high, the mass at the bottom would be too thick. Another limitation to height was that people could only practicably climb four or five stories. Because floors on the upper levels were difficult to rent, owners had no economic incentive to build any higher. But with the development of cheap methods of producing iron and steel and the invention of the elevator, architects began designing higher buildings. This spurred construction activity: landowners were motivated to develop these new buildings because the return on their investment was higher. Over time, as technological advances have allowed, the skyscraper has gotten taller.

Innovations in robotics and computer modeling have also affected the construction industry. By using computers for modeling structures and imitating wind and seismic loads, architects and engineers can better anticipate nature's constraints and create better designs to counteract them. With robots directing equipment during construction, contractors can gain more control over processes that require precision for success, such as the construction of the underwater tunnel between France and England. The desire for such projects leads to the development of technologies to make them possible, which in turn encourages similar projects to go forward.

Words, Phrases and Expressions

assembling *adj.* 组合的，装配的（combinational, assorted）
metaphor *n.* 比喻，隐喻（trope）
diverse *adj.* 不同的（various）
corporate *adj.* 公司的，法人的
oriented *adj.* 导向的
peculiar *adj.* 奇怪的（strange）
forge *vt.* 锻造，打造（establish）
intertwine *vt.* 纠缠，紧密联系（connect）
labor relation 劳资关系（relation between employer and employees）
skyscraper *n.* 摩天大楼（tower）
mass *n.* 基石（foundation block）
incentive *n.* 诱因，动机（motive）
spur *vt.* 激励（inspire）
robotics *n.* 机器人技术
model *vt.* 制作模型，建模（construct a model）
seismic load *n.* 地震力
counteract *vt.* 对抗，抵消（withstand）

疑 难 词 句

1. The organizational cultures of architects, engineers, owners, builders, manufacturers, and suppliers may seem to work against the real need to forge a partnership that will ensure the success of a project.

译文：建筑师、工程师、业主、建筑公司、制造商以及供货商的组织文化与打造一个确保项目成功的合作伙伴关系的实际需要看起来可能有些背道而驰。

2. It affects and is affected by developments in technology, computers, government policies, labor relations, and economic and political practices.

译文：建筑业影响了技术和计算机的发展、政府的政策、劳资关系以及经济与政治的惯例，同时也被这些因素所影响。

Chapter 2 The Parties To The Contracts

Most standard form building contracts are based on the assumption that the employer, the first party, who has decided to have certain works carried out for the implementation of a project, and is sponsoring the works, has decided to select a suitably qualified contractor, the second party, to execute the works.

It is also assumed that the selection of the contractor will have been made through competitive tendering based on tender documents prepared for the project by a consulting engineer.

Most building contracts cannot apply without a contract administrator (or referred to asengineer in some contracts) being appointed by the employer to administer the contract. Usually this would be the consulting engineer who has designed the project and prepared the tender documents. The contract administrator is not a party to the contract, but he plays an important role in the development process of the works. The duties that the contract administrator has to perform are defined under the contract and he must have the necessary delegated authority from the employer if he is to be able to perform them. The delegation of this authority is usually to be found in the agreement between the employer and the contract administrator.

The agreement will stipulate as the primary duty of the administrator that he carefully observes the requirements of the employer in the execution of the project. It is important to note, however, that the conditions of contract between the employer and the contractor stipulate that where, under the contract, any of the administrator's duties are discretionary, the contract administrator shall act fairly between the employer and the contractor and apply the contract in an unbiased manner. The conditions are based upon this fundamental principle and this requirement applies even if the administrator is a member of the employer's staff.

Words, Phrases and Expressions

execute *vi.* 执行，实施（to perform, implement, or carry out a scheme, a proposal, or a project)

competitive tendering 竞争性招标（selecting a contractor by means of choosing among many competing bidders)

apply *vi.* 使用，应用，适用（to fulfil, to serve its purpose, to put into use)

contract administrator 合同管理者，合同管理人员

administer *vi.* 管理，给予（to manage，to give）
consulting engineer 咨询工程师（the professional consultants advising and supervising the tendering，designing，and execution of a project）
tender document 招标文件（a request for proposal issued to the bidders who prepare and submit a tender accordingly for the sake of being awarded a project）
perform *vt.* 执行，履行（to fulfill or execute a proposal）
delegation *n.* 委托，授权（entrusting or appointing an authority or power to an organization or individual）
discretionary *adj.* 任意的，自由决定的（having one's own decision power，free to act according to one's own judgment）
unbiased *adj.* 没有偏见的（impartial）

2.1 Employer

At the end of the tendering process，the employer notifies the successful tenderer that he has been awarded the contract by issuing a letter of acceptance which records any changes to the tender documents as submitted by the contractor，resulting from agreement between the employer and the contractor，and the contract price.

The employer consents to，or declines，requests by the contractor to assign any portion of the works，prepares the contract agreement (if any) for execution by both parties，approves the performance security and the insurers as well as the terms of the insurance policies submitted by the contractor. The employer will wish to ensure that the contract works insurance is in accordance with the laws and regulations of the country in which the works are to be executed and that the policy adequately covers the employer's risks and the deductible limits are acceptable. Provided it is acceptable to the employer the contractor will normally use his customary sources for the provision of securities and insurance.

The Employer makes the advance payment (if any) against a suitable guarantee from the contractor and authorises the contractor to move on to the site. During the period of the contract the employer makes payments to the contractor as certified by the contract administrator to be due under the contract.

The employer takes over sections of the works as they reach substantial completion，if this is required under the contract，and ultimately takes over the whole of the works following the issue of certificates by the administrator. In the event of the contractor becoming liable for liquidated damages，the employer may deduct an amount in accordance with the conditions of contract.

The employer may authorise work to be completed by others if the contractor is in default. The employer can terminate the contract in the event of the contractor failing to perform or in certain other circumstances defined in, and subject to , the law governing to the contract. The employer, if he defaults, can also be subject to cancellation of the contract by the contractor or to suspension of work by the contractor.

The employer and the contract administrator should maintain such contact with each other as will facilitate smooth and unhindered progress of the works. The employer should respond, without delay, on all matters for which the administrator is required by the contract to consult the employer before issuing an instruction, determining an amount to be added to or deducted from the contract price or granting an extension of time.

Words, Phrases and Expressions

tenderer *n.* 投标者 (a bidder who prepares and submits a tender)
letter of acceptance 中标函 (a letter accepting the tender)
performance security 履约担保 (a guarantee to ensure the contractor well fulfil the contract)
insurance policy 保险单 (a written statement of the terms of a contract of insurance)
deductible limit 免赔额 (an insurance term meaning a sum which the insured must pay by himself when settling a claim)
advance payment 预付款 (a sum of money the employer pays to the contractor at the outset of a project as a loan for the contractor to mobilize the execution of the works)
substantial completion 基本竣工 (practical completion which enables the employer to issue the taking-over certificate to the contractor)
liquidated damages 误期损害赔偿金 (a penalty which is paid by the contractor to the employer when the time for completion is delayed by the contractor's default)
conditions of contract 合同条款 (terms or provisions of a contract)
default *n.* 违约 (breach of contract)
suspension *n.* 暂停 (stop for the time being)
facilitate *vi.* 使便利，有助于 (to help)
extension of time 延长工期 (extending the time for completion)

疑难词句

1. The employer will wish to ensure that the contract works insurance is in accordance with the laws and regulations of the country in which the works are to be executed and that the policy adequately covers the employer's risks and the deductible limits are acceptable.

译文：业主将希望能确保合同工程的保险符合工程所在国的法律和规定，保险单能足以涵盖业主的风险，并且免赔额也能够被接受。

2. The employer should respond, without delay, on all matters for which the adminis-

trator is required by the contract to consult the employer before issuing an instruction, determining an amount to be added to or deducted from the contract price or granting an extension of time.

译文：业主应该对合同要求合同管理者发出指令前必须与业主商量的所有事宜，以及对合同价的增减的确定，或给予延长工期的决定等事宜立即答复，不得延误。

Notes：

1. **substantial completion** 基本竣工 This term stands for a practical or basic completion of the works under a contract. In this case, the contractor has completed the entire works in the way that all the functions of the works have been completed to the specifications of the contract and the employer can take over and use it. This so-called substantial completion, however, is different from 'absolute completion' or 'full completion' in that the works of substantial completion leaves some trivial or slight outstanding work unfinished which does not hinder the employer from using the works.
2. **extension of time** 延长工期 This term means extension of the time for completion which is granted to the contractor by the employer or the contract administrator on behalf of the employer as a result from occurrence of any events which has affected the contractor's process of executing the works provided that such events are not caused by the contractor or not because of the contractor's default.

2.2 Contractor

The obligation of the contractor is to execute and complete the works, for which he has submitted his tender, within the time specified in the contract. In addition he has obligation to remedy any defects which appear during the defects liability period.

As soon as is reasonably possible after receiving notification from the contract administrator, the contractor shall submit the securities, guarantees and insurance policies required by the contract and shall commence the works. He prepares the construction programme, provides all necessary materials, contractor's equipment, temporary works, management, superintendence and labour and selects the method of carrying out the works. The contractor is not responsible for the design and specification of the permanent works unless expressly provided for in the contract nor for any temporary works not designed by him.

The contractor receives and complies with instructions from the contract administrator acting on behalf of the employer and is responsible for the care of the works throughout the construction period until the works are officially taken over by the employer or are deemed to be taken over by the employer.

The contractor is responsible for his own staff and work force and for taking out social and other insurances in respect of his personnel. He must comply with all applicable laws, by-laws and regulations and ensure that all those for whom he is responsible also comply.

Under normal circumstances the contractor designs all temporary works and submits his proposals, with supporting calculations, to the contract administrator for comment. If, during the construction period, he encounters unforeseen physical obstructions or conditions on the site he notifies the administrator who issues relevant instructions. The contract administrator will review the circumstances and after consultation with both parties will determine to what extent, if any, the contractor may be reimbursed for additional costs or granted an extension of the time for completion.

In the event of default by the employer the contractor may suspend progress of the works or reduce the rate of work and claim an appropriate extension of time and/or additional payment.

Normally there will be one main or principal contractor who signs the contract and has overall responsibility for the execution and completion of the project. There will usually be a number of subcontractors working on the site undertaking specialist contracting activities. The subcontractors are responsible to the contractor for material, workmanship, performance and progress and the contractor is responsible under the contract for each subcontractor's work and behavior.

On occasions the employer will wish to have a particular subcontractor engaged because of his knowledge of that subcontractor's skills or because of his knowledge of some process, materials or plant particularly required by the employer. A subcontractor selected in this way is known as a nominated subcontractor. Once he has been accepted by the main contractor the latter is responsible for the work of the nominated subcontractor to the same extent as applies to the work of all other subcontractors. For this reason it is important that nomination is not misused by the employer. Failure to perform satisfactorily by a nominated subcontractor can cause many difficulties on site. The particular conditions applying to the appointment of nominated subcontractors are dealt with in the relevant clause of the conditions of contract. The contractor may object to the nomination for good and sufficient reasons (e. g. lack of experience or financial strength) and engagement against the wishes of the contractor will probably disturb harmony on the site. There would have to be very exceptional reasons for such a thing to happen.

In very large or complex projects a number of contractors may form a joint venture to act as the contractor. In such cases the same principles apply as in the situation with only one contractor. In the case of a joint venture the employer would normally require that all the parties to the joint venture have joint and several liability.

In projects where several contractors are operating on a single site under individual contracts, each contractor must give the other contractors reasonable cooperation and opportunities for carrying out their work and this should be reflected in terms of the contracts and in the respective programmes.

Words, Phrases and Expressions

remedy *vt.* 修复（to make good, put right）
programme *n.* 计划，安排（schedule, scheme）
temporary works 临时设施（site facilities used for execution of a project which are to be removed after the project is completed, such as site office, dormitory, dining hall, batching plant, etc.）
specification *n.* 技术要求，技术规范（technical requirement the workmanship of the works, including the works' materials, permanent plant, parts, and etc.）
expressly *adv.* 清楚地，明白地（clearly, explicitly, opposite to 'impliedly'）
take over 接收（to receive a project when it is substantially completed）
comply with 服从，遵从（obey）
review *vt.* 审查，回顾（examine）
reimburse *vt.* 偿还（repay）
main or principal contractor 总承包商（a general contractor who signs an agreement with the employer directly and who may sublet parts of the works to a subcontractor）
subcontractors *n.* 分包商（a specialty contractor who subcontracts a part of a project from a main contractor）
nominated subcontractor 指定分包商（a subcontractor who is appointed by the employer directly）
engagement *n.* 约定，雇佣（employment）
joint venture 联营体（a consortium composed of several firms to tender a project as one bidder）

疑 难 词 句

1. The contractor may object to the nomination for good and sufficient reasons (e. g. lack of experience or financial strength) and engagement against the wishes of the contractor will probably disturb harmony on the site.

译文：承包商可以凭合理充足的理由（如缺乏经验或财力）反对该指定，无视承包商的意愿的雇佣可能会扰乱现场的和谐。

Notes:

1. **programme** 施工计划 This word stands for a work schedule which the contractor shall submit to the contract administrator within a specified period after receiving the notice of commencement. This programme should be updated by the contractor whenever it is inconsistent with the actual progress.

2. **temporary works** 临时设施 This expression means all temporary works of every kind required on site for the execution and completion of the permanent works and the remedying of any defects, such as site office, dormitory, dining hall, batching plant, warehouse, and etc. which should be removed when the permanent works is completed or when they are no longer used at site.

2.3 Contract Administrator

The contract administrator (also referred to as engineer in some contracts) is not a party to the contract between the employer and the contractor but his terms of engagement are set out in an agreement between the employer (client) and the contract administrator (engineer or consultant).

The duties under most standard form contracts which are allocated to the contract administrator (Engineer) include the issue of information and instructions to the contractor as the work proceeds, commenting on the contractor's proposals for carrying out the work, ensuring that materials and workmanship are as specified, agreeing to measurements of work done and checking and issuing to the employer interim and final payment certificates. In administration of the contract all communications with the contractor pass through the contract administrator, thus avoiding possible confusion and misunderstanding although meetings between the employer, the contractor and the administrator should be held regularly. The contract administrator's duties will normally include instructions relating to management of the contract and changes in the nature and extent of the work, the cost thereof and the time for completion. For example, the issue of instructions to proceed with or to suspend the progress of the works is a matter of management. In order to maintain confidence between the parties it is important that the contract properly discloses the procedures for the contract administrator's action in such matters and that the administrator duly observes such procedures.

Many of the functions allocated to the contract administrator involve financial matters. The certificates he issues for interim and final payments include all cost elements arising under the terms of the contract other than those (if any) resulting from arbitration proceedings and the application of the liquidated damages clause. Under normal circumstances the payments certified by the administrator will lie within the contract price and will, therefore, have already been authorised by the employer, but variations ordered under the relevant clause, fluctuations in the price of labour, materials and freight under the contract, works done or materials and services supplied and the settlement of the contractor's request for additional payment under the contract could each result in an adjustment to the contract price.

Many building contracts, for example the FIDIC Conditions, are based upon the principle that the engineer has the authority to determine additional payments. This is in the interests of efficient management and avoidance of duplication of effort. It also ensures that the tender price from efficient contractors will be lower than would be the case if the contractor did not feel sure that when he was required to do additional work his remuneration for such work would be evaluated by a professional able to judge the value of such work. In the event the employer wishes to limit the authority of the administrator this should be clearly stated in the contract so that the contractor is aware in advance of tendering the conditions under which he is required to work. The degree to which the employer leaves the contract administrator to determine matters affecting the extent and cost of the works and the time for their completion will depend to a large extent on the in-house capability of the employer. Therefore, the contract administrator should be selected having regard to his professional integrity and his ability to fulfil his obligations under the contract and to administer the contract fairly and impartially in the interests of both parties.

In the exercise of his duties the administrator accepts the responsibilities attached to them. These responsibilities should be clearly defined in the agreement between the employer and the contract administrator and should be made known to the contractor.

As the works progress the contract administrator will be required by the contract to give instructions, give or refuse approval or consent, approve work, authorise payments, issue certificates, etc. It should be understood by both parties to the contract that in giving approval or consent and such other acts which are the duty of the administrator his objective is to ensure that the employer receives the works at completion in accordance with the requirements of the contract and that the contractor is suitably rewarded for the work he carries out.

Words, Phrases and Expressions

allocate *vt.* 分配，分派（assign，distribute）

interim payment certificate 期中支付证书（the temporary or monthly payment certificate issued by the contract administrator or engineer to the employer for making payment to the contractor for the work done）

final payment certificate 最终支付证书（the last payment certificate following the settlement between the employer and contractor，which is issued after the engineer receives the final statement and written discharge）

administration *n.* 管理（management）

communication *n.* 沟通，交流（exchange）

regularly *adv.* 定期地，经常地，不间断地（frequently，continuously）

procedure *n.* 程序，步骤

arbitration proceedings 仲裁程序

liquidated damages 误期损害赔偿（a penalty on the part of the contractor who fails to complete a project within the time limit specified by the contract）
variation *n.* 变更（a change in the technical specification, work scope or nature of the design, fabrication, and all the works originally specified by a contract）
fluctuation *n.* 波动，起伏（moving up and down）
settlement *n.* 结算，解决
duplication *n.* 重复（repetition）
remuneration *n.* 报酬（reward）
evaluate *n.* 估价（value）
in-house *adj.* 组织内部的（inner, within, itself）
professional integrity 职业道德
impartially *adv.* 公正地（fairly）
exercise *n.* 行使，运用

疑 难 词 句

1. The duties under most standard form contracts which are allocated to the contract administrator（Engineer）include the issue of information and instructions to the contractor as the work proceeds, commenting on the contractor's proposals for carrying out the work, ensuring that materials and workmanship are as specified, agreeing to measurements of work done and checking and issuing to the employer interim and final payment certificates.

译文：大多数合同划定给合同管理者（工程师）的职责包括：工程进行中向承包商发布信息和指示；评价承包商的施工方案，保证材料和工艺符合规定；同意已完成的工作的测量值；校核并向业主送交期中和最终支付证书。

2. Under normal circumstances the payments certified by the administrator will lie within the contract price and will, therefore, have already been authorised by the employer, but variations ordered under the relevant clause, fluctuations in the price of labour, materials and freight under the contract, works done or materials and services supplied and the settlement of the contractor's request for additional payment under the contract could each result in an adjustment to the contract price.

译文：正常情况下，合同管理者开具证书的付款将包括在合同价格中，因此这些款项将是已经获得业主授权的。但根据相关条款指令的变更，合同允许的劳务、材料和运费价格的波动，以及承包商根据合同提出增加付款的已完工程或已提供的材料和服务的每一项都可以导致合同价格的调整。

3. It also ensures that the tender price from efficient contractors will be lower than would be the case if the contractor did not feel sure that when he was required to do additional work his remuneration for such work would be evaluated by a professional able to judge the value of such work.

译文：同时它还能确保高效率的承包商的投标价格低于下列情况时的价格，即当承包商被要求从事附加工作时，他不能确信，其从事该工作应得的报酬是否由一位有能力判定该工作价值的专业人员进行估价。

Notes：

1. **Interim payment certificate** 期中支付证书

   Unlike manufacture of an ordinary product，a project of construction of a building will cost a lot of money and a long period of time. Therefore，the contractor cannot afford to wait to be paid after the whole works is completed and handed over to the employer. In order to solve this problem，most building contracts entitle the contractor to be paid by instalments for the works done within a regular time，for example every month. Thus，during the process of construction，the employer makes payment to the contractor every month for the works completed by the contractor in this month. To do so，the contractor should apply for this monthly payment by submitting a monthly statement to the contract administrator or engineer for review and certification，then the contract administrator will issue an interim payment certificate to the employer who will then make payment to the contractor accordingly.

2. **Final payment certificate** 最终支付证书

   The final payment certificate is different from the interim payment certificate in that the final payment certificate is a final settlement between the two parties，which states the amount which is finally due，it，therefore，is a balance due after the offset of all amounts previously paid by the employer and all sums to which the employer is entitled.

Chapter 3 Procurement Methods of Construction Contracts

In the construction industry, unlike other industries, there are many different procurement methods which can be adopted to contract a construction project. Those procurement methods respectively apply to different natures of project as well as the different employer's requirements. The major procurement methods of contraction contracts are discussed below.

3.1 General Contracting

General contracting involves the separation of construction from design. A main contractor is employed to build what the designers have specified. Since this form of procurement was developed it has become very common and is often referred to as traditional. Although the popularity of general contracting is waning nowadays, it still accounts for most of the construction work in the world. However, some of the basic defining characteristics of general contracting no longer hold true.

3.1.1 Background

Traditional general contracting has been around over 100 years, before that time construction work tended to be procured by a series of direct contracts between client and trade contractors. General contracting was a response to the increasing sophistication of construction technology during the Industrial Revolution. As more and more techniques and materials increased largely, co-ordination problems on building sites became more complex. Meanwhile, a number of issues such as continuity of employment of suitable participants, wide application of plant, the development of the transport infrastructure, make the idea of general contracting a viable procurement method. The desire of administrators to focus more on design and client-related issues and less on the day-to-day business of construction fuelled demand for a contractor who would shoulder all of the responsibility for building the project.

The continuing development of science and technology has led to the emergence of many new specialist skills. Thus, a consequence of the increased use of specialists is a corresponding increase in the demand for coordination and integration.

As a result, integration and co-ordination could take place by means of participants making mutual adjustments in their own work. In the modern technological building process, very few people can grasp all of the different technologies. The craftsmen cannot be left to sort out the relations between work packages; these become the responsibility of the designer. The result is an increased demand for management.

The basic defining characteristic of general contracting is that the contractor agrees to produce what has been specified in the documents. Designers, on behalf of the employer, produce the documents and builders produce the building. In theory, the contractor should be invited to price a complete set of documents that describe the proposed building fully. Such documentation demands that the architect (or lead designer) co-ordinates design advice from a wide variety of specialists. The result is that the contractor has no responsibility for design. The contractor's offer of price is based on the bill of quantities, a document that itemizes and quantifies, as far as possible, every aspect of the work. The bill forms not only the pricing document but also, because of its comprehensiveness, an important mechanism for controlling.

Words, Phrases and Expressions

trade contractor 专业承包商，分包商（specialist contractor, sub-contractor）
sophistication *n.* 复杂性（complexity）
infrastructure *n.* 基础设施（public service facilities, such as highway, telecommunication, railway, bridge, ports, airports, etc.）
viable *adj.* 切实可行的，可操作的，可实现的（realistic, workable, practical）
integration *n.* 结合，一体化，整体性（combination, incorporation）
sort out *vt.* 整理，处理（clear up, deal with）
work package *n.* 工作包，工作单元（major piece divided from the work scope of a project）
itemize *vi.* 逐项列出（to list item by item）
quantify *vt.* 定量
comprehensiveness *n.* 广泛，综合（wideness, being embracing）
mechanism *n.* 办法，途径，机制（means, approach）

疑 难 词 句

1. **The desire of administrators to focus more on design and client-related issues and less on the day-to-day business of construction fuelled demand for a contractor who would shoulder all of the responsibility for building the project.**

译文：合同管理者对设计以及与业主相关的事宜，比对施工方面的日常事物更注重的愿望，激发了对能够承担项目建造的全部责任的承包商的需求。

Notes: This sentence shows a reason causing the tendency toward the procurement method of general contracting because the general contracting requires the contractor to be able to administer the whole process and all the aspects for building the project and coordinate all the participants in their own roles.

2. **Thus, a consequence of the increased use of specialists is a corresponding increase in the demand for coordination and integration.**

译文：这样，更多（更广泛）地使用专家的结果就导致了对协调和整体性的需求有了相应的增加。

Notes：This sentence indicates that the requirement of coordination and integration is a key circumstance in which general contracting is typically used.

3.1.2 Basic Characteristics of Traditional General Contracts

As mentioned above, the key feature is the involvement of the architect in a design and co-ordination role, and the contractor in a production role. This means that the contractor is not, theoretically, liable for design, but only for workmanship. In practice, however, the division between design and workmanship is not always clear-cut. The design must be documented in order for the contractor to be able to tender for the work. However, the documentation will not cover everything. Many detailed aspects are within the skill and knowledge of a competent contractor. Therefore, the exact position of individual nails in a floor, or the location of joints between different pours of concrete and their formation, are matters which are often left to the contractor. There are also many factors which are not documented simply because no-one thought of them before the site work started. In these cases, it is essential for the contractor to seek clarification from the architect. A contractor who makes assumptions (even when based upon common sense and experience) will incur design liability for those choices. This leaves the employer exposed because the contractor will probably not carry indemnity insurance for such design decisions.

This leads to the next distinctive feature, which is that the contractor undertakes to do the work described in the documents. It is rare to find a building contract which places upon the contractor the obligation to supply a particular building. But general contracting usually obliges the contractor to carry out the works shown and described in the contract documents. The contracts also state that the quality and the quantity of the work is that which is described in the contract. Bills of quantity, if used, will usually be required by the contract to be prepared according to the relevant standard method of measurement. In building contracts, certain items specified as such in the bills, have to be done to the reasonable satisfaction of the architect, whereas in civil engineering contracts the contractor must ensure that all work is in accordance with the contract, to the satisfaction of the engineer. Some contracts go as far as saying that the contractor must comply with any instruction of the contract administrator, on any matter concerning the works whether mentioned in the contract or not. Clearly, there is a wide variety of practices in the way that standard contracts deal with the contractor's obligations.

A critical feature of general contracting is the way in which nomination of sub-contractors is used. It is an excellent mechanism for ensuring that the contractor employs sub-contractors of adequate standing. Without nomination, contractors will use high calibre sub-contractors when tendering, in order to keep the bid price high, but on winning the job they frequently re-negotiate to find the cheapest possible sub-contractor, in order to maximize the return on the contract. Nomination forces the contractor to use a particular subcontractor. However, the abuses of this mechanism are frequent. One typical 'abuse' is where nomination is used to relieve the architect of the burden of having to detail the design and specification fully. By nominating a sub-contractor, the design team can get away with minimal documentation. However, this is bad practice, and it will generate problems during the construction stage. If everything is measured and/or specified adequately, then most of the problems can be overcome.

The power to change the specification, known as a variation, is a feature of general contracts. This gives the contract administrator the power to change the work required of the contractor. There are usually detailed provisions for valuing the financial effect of variations, and these are based, as far as possible, upon the contractor's original price. The principle here is that the contractor should be paid according to what would have been included in the bills, had the contractor known about the varied work at the time of tendering. Only in exceptional circumstances should the basis of the payment to the contractor be total cost reimbursement (day-work). Variations clauses enable the employer's design team to refine the design as the contract progresses, but the provisions are often abused by careless employers who see the opportunity to make arbitrary changes to the works as they proceed. This practice leaves the client dangerously exposed to claims from wily contractors who can demonstrate all sorts of consequent effects, which would attract extra payment under the contract.

Payment provisions are typically based upon the assumption that the contractor will be paid in instalments for the work. Under most forms these instalments are based upon the value of the work executed to date and are regulated by the use of certificates. Interim certificates will state how much money is due to the contractor. Other certificates may be issued to record non-completion, completion, making good defects and so on. All certificates are issued by the contract administrator. A small amount of money on each interim certificate is retained by the employer in a retention fund. This fund is used to encourage the contractor to remedy any defects in the contract after completion.

Completion is rarely a clear-cut affair. When the work is substantially complete, a certificate is issued, known variously as a certificate of substantial completion, certificate of practical completion or taking-over certificate. At this point a period commences during which time certain defects may be the responsibility of the contractor. This period is variously known as defectsliability period, defects correction period or maintenance period. This period is usually six or twelve months. During this period, the contractor has the right to remedy any defects which become apparent in the completed building. Without such a right, the employer would be entitled to employ other people, and charge the contractor for the work.

The contract period may be extended for various specific reasons. Such an extension will relieve the contractor of liability for liquidated damages for that particular period. Liquidated damages are an estimate of the employer's likely losses which may be incurred by late completion. An amount of money is stated in the appendix to the contract. This amount has to be estimated before the contract is executed. It is not necessary for the employer to provide evidence of such financial loss, because the amount of money levied against the contractor is not related to actual loss; only to the loss which could have been reasonably foreseen at the time of tender. If the amount of money is not a realistic pre-estimate, or if it is punitive, then it is deemed to be a penalty.

Another financial provision is that related to changes in market prices of the contractor's resources. These may be dealt with in a fluctuations clause. On a contract with a short duration, it is unlikely that there would be much alteration in the price of the contractor's supplies. However, in periods of high inflation, or on long contracts, the risks of price rises in such resources are very high. It would be unwise to impose such a risk on the contractor. It is more economical to the employer to absorb these risks, and therefore a fluctuations clause can be used which effects changes to the contract sum based upon the market prices for labour, materials and other contractors' costs.

In general contracting, the designers act on behalf of the employer in converting the employer's requirements first into a brief and subsequently into a workable design. In such a case, the lead designer has much flexibility in choosing the design team. Therefore the lead designer demands good co-ordination and integration skills, as well as good communication skills. In addition, the lead designer must develop a clear understanding of the employer's requirements. Therefore, the lead designer demands a wide range of skills.

Also in general contracting, in some circumstances, the employer may wish to nominate a sub-contractor with whom the main contractor must enter into a contract. By this way, the employer may control over the election of certain specialist sub-contractors to ensure the chosen sub-contractor to have a proven track record for good work, and to use a sub-contractor with whom the employer might have developed a long-term business relationship and also to base the selection of the sub-contractor on a basis other than lowest bid. In such a case, a contractor will be paid whatever nominated sub-contract work costs. Such work falls into the category of cost reimbursement, unlike the case with the contractor-selected sub-contractor where the contractor pays them from within the contract price.

Words, Phrases and Expressions

workmanship *n.* 技艺，工艺（skill shown in a finished craft product, craftsmanship）
clarification *n.* 澄清（making clear or understandable, explaining）
assumption *n.* 假设（to suppose, to accept or believe to be true）
standard contracts 标准格式合同（format contract）
adequate standing *n.* 良好的名望，声誉，地位（high reputation）
caliber *n.* 才干，能力（competence）
bid price 投标价（offered price）
nomination *n.* 指定，任命（appointment, designation）
cost reimbursement *n.* 偿还，退还（repaying expenses）
arbitrary *adj.* 随意的，独断的，专横的（decided by chance or whim, imperious）
demonstrate *vt.* 证明（to prove）
installment *n.* 分期付款（divided payment）
retention fund 保留金（a part of payment withheld by the employer for repairing the defects which might be found later）
defect *n.* 瑕疵，过失（flaw, mistake）
defects liability period 缺陷责任期（maintenance period during which the contractor is liable to make good the defects of the works）
relieve *vt.* 消除，解除（to release, to remove）
fluctuations clause 价格波动条款（escalation clause）
inflation *n.* 通货膨胀（price rise）
brief *n.* 诉讼事实摘要，诉讼事件；（律师的）辩护状

疑 难 词 句

1. This leaves the employer exposed because the contractor will probably not carry indemnity insurance for such design decisions.

译文：这会使业主非常不利，因为承包商可能不会为这些设计决定投保障险。

Notes：exposed 意思是 being exposed to unfavourable things or conditions，可译成“处于非常不利的情况”。

2. It is an excellent mechanism for ensuring that the contractor employs sub-contractors of adequate standing. Without nomination, contractors will use high calibre sub-contractors when tendering, in order to keep the bid price high, but on winning the job they frequently re-negotiate to find the cheapest possible sub-contractor, in order to maximize the return on the contract.

译文：这是一个很好的机制，它可以确保承包商雇佣具备良好声誉的分包商。如果没有这种指定，承包商可能会在投标时用能力强的分包商来将投标价提高，而一旦赢得了项目，为了利润的最大化，他们又往往会重新谈判去另外寻找尽可能便宜的分包商。

3. The principle here is that the contractor should be paid according to what would have been included in the bills, had the contractor known about the varied work at the time of tendering.

译文：这里有一个原则：如果承包商在投标时就知道了这些变更，承包商就只能得到工程量清单已包含的各项支付（而不能得到额外的支付）。

Notes: It means (or implies) that the contractor will not get any additional payment for the variation unless these varied work was not known to the contractor at the tender stage.

4. During this period, the contractor has the right to remedy any defects which become apparent in the completed building. Without such a right, the employer would be entitled to employ other people, and charge the contractor for the work.

译文：在此期间，承包商有权对已完工程中出现的瑕疵进行修复。如果没有这个权利，业主就有权雇其他人来修复这些缺陷，然后向承包商索取该费用。

Notes: Usually, under most standard form contracts, the employer is not entitled to employ other firms to repair the defects during the defects liability period, unless the contractor, after being notified thereof, fails to do it within the time specified in the contract.

5. In such a case, a contractor will be paid whatever nominated sub-contract work costs. Such work falls into the category of cost reimbursement, unlike the case with the contractor-selected sub-contractor where the contractor pays them from within the contract price.

译文：在此情况下，业主将承担其指定分包商的工程费用。这些工程属于成本补偿的范围，而不像承包商自己选定的分包商那样由承包商从合同价中支付费用。

Notes: Usually, the works to be done by the nominated sub-contractor is not included in the tender submitted by the bidder. Therefore, when the bidder is awarded the project, the contract price will of course not include the works to be completed by the nominated sub-contractor.

3.2 Design and build

"Design and build" is a procurement method which has been in use for a long time. Before the architecture became a profession separate from construction, the users would procure buildings by a process of design and build. It was the separation of responsibility for construction from the responsibility for design that led to the emergence of so-called traditional general contracting in the nineteenth century. But this separation of design from construction in the building industry has, for a long time, been the source of many problems.

3.2.1 Background

As the roles of the professional consultants continue to develop and change, the methods of procuring buildings also change. Because the contractual relationships in the construction process are difficult and complex, most people who participate in the process prefer to standardize the contracts. Design and build is one type of variety of 'package deals' in construction, and it is clear that this procurement method has a very wide applicability.

3.2.2 Features of Design and Build Contracts

There are some essential features unique to design and build contracts. These features can best be dealt with in terms of how the employer describes the requirements for the job; how the contractor proposes to achieve them; the pricing mechanism; and the roles and responsibilities within the process.

Employer's requirements and contractor's proposals

The first of the essential features of a design and build contract is that the employer goes to a contractor with a set of requirements. The contractor prepares proposals accordingly which will include construction as well as design work. The contractor's design input varies from one contract to another, ranging from the mere detailing of a fairly comprehensive design to a full design process including proposals, sketch schemes and construction information. There will usually be some negotiation between the employer and the contractor, the aim being to settle on an agreed set of contractor's proposals. These proposals will include the contract price and how it has been calculated.

Once the contractor's proposals are responsive to the employer's requirements, the contract can be executed and the contractor can carry out the work. Thus, the contractor will be totally responsible for undertaking the design work outlined in the contractor's proposals, for building the building, and for coordinating and integrating the entire process. This includes the appointment of consultants if the contractor is not able to do the job by himself. The employer may also choose to appoint consultants in order to monitor the various aspects of the work, but this in not always the case.

Price

Another feature of design and build contracts, although which is not always the case, is a guaranteed maximum price (GMP). This helps clients to feel reassured that they are not signing a blank cheque! As a motivation to the contractor, any savings made by completing the project for a price below the GMP may sometimes be shared between the client and the contractor.

The price for design and build arrangements is controlled by means of a contract sum analysis. There are no bills of quantities needed. The nature of this pricing is very different to that of a bill of quantities, its form not being specified by the contract. It can be presented in any form which is proper for the situations of the project, but sometimes with a similar aim of bills of quantities. These are calculation of stage payments, valuation of client's change order, and implementing the escalation clauses, and etc. This pricing by contract sum analysis should help the design content of the work to be reflected in any estimation of the value of variations.

Roles and responsibilities

One of the most significant features of design and build deals is the lack of an independent certification role in the contract. For example, there is no architect or contract administrator to dissolve the possible disputes between the parties and there is no independent quantity surveyor responsible for preparing the basis on which contractors bid.

In the case of design and build deals, many traditional quantity surveyors have become specialist client advisers in all aspects of the economics of building and the procurement process. There are a large number of consultants who are qualified to advise the client during the whole of the process. These consultants would be used by the client when dealing with the contractor. And very often in the case of the design and build contracts, the well-informed client with the experience of the traditional general contracting will set up their own project team, with a counterpart for each of the professional advisers in the contractor's project team for a better control over this procurement process of design and build.

Advantages and disadvantages

The contractual relationships in design and build offer some advantages over other methods of construction procurement. The most important advantage is that the contractor is responsible for everything. This "single-point" responsibility is very favorable to clients, because it enables clients to avoid the troubles to distinguish the difference between a design fault and a workmanship fault. This single-point responsibility also means that the contractor is not relying on other firms (e. g. architects) for the design work or for other works associated with the contract. Besides, by more effective communication due to the single-point responsibility, experience has shown that programmes and budgets are more likely to be fulfilled, and the speed of construction is likely to be quicker.

The design and build process increases opportunities for utilizing the benefit of the contractor's experience during the design stages of the project. As the contractor is responsible for the design work, the so called "buildability" can be realized. And the benefits of the integration of designers and builders are more economic buildings as well as a more economic and effective production process.

One possible disadvantage of design and build is where there is a conflict between aesthetic quality and ease of construction; the requirements for fabrication will dominate. A further criticism has been that a design and build contractor will put in the minimum design effort required to win the contract. These two criticisms suggest strongly that quality, particularly architectural quality, will suffer under this procurement process. However, this is not a valid criticism of the process itself, but rather of some of the people who may be exercising it. As such, it is a criticism which can be made about any process.

Unlike some methods of procurement, it is not necessary for the design and build client to be an 'expert' client. And cost certainty for the employer is one of the advantages of design and build. It is a fixed price contract, so the risk associated with pricing is entirely the contractor's, except to the extent that fluctuations clauses apply. However, notwithstanding the cost certainty, design and build is not necessarily cheap. Single-point responsibility, and fixed price contracts, mean that the contractor carries more of the risk than a general contractor would. Risk attracts a premium, so it is to be expected that a design and build contractor would add a premium to a tender to allow for this extra risk. In this respect, design and build may be more expensive than traditional general contracting. On a high-risk project it may be inappropriate to pass too much risk over to the contractor so other forms of procurement should be considered.

An element of competition is usually seen as advantageous when appointing a contractor. Traditionally, contractors submit tenders based upon a description of the works. In this situation they are competing on price alone, since they are all pricing the same description of the works. One of the strengths of design and build is that the contractor's proposals will include design solutions to problems posed in the employer's requirements. Here contractors are competing not only on price but also on any other criteria which the client thinks important. This presents opportunities for making the level of accommodation the selection criteria, where a client can put forward a budget and ask the bidders to demonstrate how much building they can supply for the money. Clearly, the element of competition is far more flexible in design and build than in traditional arrangements.

One of the biggest disadvantages of design and build is that the brief (in terms of the employer's requirements) must be very clear and unambiguous. Also, it should not be subject to change during the project. This is not such a problem in general contracting because the brief can be developed alongside feasibility study and sketch schemes. Therefore a client who wishes to reserve the right to alter requirements during the fabrication process should not use design and build. The limited scope for variations and changes is thus a weakness of the design and build process.

Single-point responsibility is the most obvious advantage offered by design and build. There is no division of responsibility between design and fabrication, so the finished building should reflect the trade-offs made between the design exigencies and the fabrication exigencies. It is a method to be used where the project is one for which it makes sense to combine responsibility for design with responsibility for fabrication. The types of project for which design and build contracts are suitable are those where the contractor's responsibility for design extends over the whole project, whether the design is partially completed by others or not. This is as distinct from the situation where the contractor's responsibility for design only covers a particular portion of the works.

Since the contractor is undertaking the design work, there are opportunities to overlap the design and construction processes and thus to make an early start on site.

The risks in design and build are almost all on the part of the contractor, for instance, the risk of the cost exceeding the contract price lies entirely with the contractor; any fault in the finished building will be the liability of the contractor, because it is not necessary for the employer to attempt to distinguish whether a particular problem is a design fault, a manufacture fault or an assembly fault. Again, single-point responsibility means that the contractor is responsible for ensuring that the project is completed on time. Any delays beyond the control of the employer would be at the risk of the contractor in the absence of express provisions to the contrary. And quality is not compromised either simply by using the design and build form of procurement.

Words, Phrases and Expressions

package deal 一揽子交易（a set of, a series of, business）
input *n.* 投入（the work, supplies, or effort put in）
carry out *vi.* 执行，贯彻，进行，实施（to execute, implement, achieve）
responsive *adj.* 符合的，响应的（responding to, consistent）
motivation *n.* 动机，动力，诱因（incentive, inducement）
escalation clause 调价条款（fluctuation clause）
certification *n.* 证明，证实（testifying, approving）
dispute *n.* 争端，纠纷（dissension）
quantity surveyor 物料估算师（cost estimator）
consultant *n.* 咨询工程师，顾问（advisor）
single-point responsibility 单点责任（一方承担全部责任，避免了互相推诿）
buildability *n.* 可建造性（being possible to be fabricated）
premium *n.* 报酬，奖金，溢价（reward, prize）
accommodation *n.* 住处，提供居住的空间或容纳能力（lodgings, space, capacity to receive people）
fabrication *n.* 制作，构成（manufacture, construction）
trade-off *n.* 权衡，平衡，交换（balance, weighing）
overlap *vi.* 重叠（to lap over）
express provision 明示条款，明文规定（a provision explicitly stated out in a contract, opposite to implied provision）

疑 难 词 句

1. This pricing by contract sum analysis should help the design content of the work to

be reflected in any estimation of the value of variations.

译文：这种通过对合同价的分析来定价的方法，应该能帮助在变更估价中反映出设计工作所占的比重。

Notes：Usually，without the contract sum analysis，when the varied work is valued，the design work in the variation is hard to be valued especially in the case of general contracting contract. However，under the design and build contract，this problem is easily solved by the contract sum analysis.

2. And very often in the case of the design and build contracts，the well-informed client with the experience of the traditional general contracting will set up their own project team，with a counterpart for each of the professional advisers in the contractor's project team for a better control over this procurement process of design and build.

译文：在设计与建造合同中，熟悉这种承包方式并且有传统的总承包经验的业主，经常会成立一个拥有与承包商项目部中的专业人员对等角色的自己的项目部，以便更好地控制这种设计与建造一体的承包方式。

3. And the benefits of the integration of designers and builders are more economic buildings as well as a more economic and effective production process.

译文：这种设计人员和施工人员一体化的承包方式不仅使我们的生产过程更加经济和高效，而且还能使我们建造出更具有经济效益的建筑物。

Notes：Under this type of design and build contracting，the designers will consider more the ease of the consequent fabrication of the building as well as the ease of the usage of the building by the client，and meanwhile the designers will bring some economic benefits into their scheme because the potential contractor of a design and build contracting project has to compete with others to win the project.

4. This presents opportunities for making the level of accommodation the selection criteria，where a client can put forward a budget and ask the bidders to demonstrate how much building they can supply for the money.

译文：这就使我们有机会将住房的容纳能力作为选择承包商的标准，如业主可以提出一个预算额，让投标商论证他能用这笔钱提供多少住房。

Notes：This is a typical example for a flexible competition criteria for selecting contractors which only the design and build procurement method can realize，because the other procurement methods，taking the traditional general contracting as an example，can only ask the bidders to compete on price alone.

Notes：

1. **Guaranteed maximum price** 最高保证价格，价格上限

Using a design and build contract，the employer assumes the risk of greater contract price than other forms like general contracting and etc. Therefore the employer may desire to set a top limit on this risk by requiring the contractor to guarantee the maximum cost of the contract and thereby fix a limit on the amount of the employer's investment in the project. This guaranteed maximum price can be used along with a savings clause, then the contractor is paid upon completion of the contract a certain percentage of the amount by which the actual cost of performance falls below the guaranteed maximum cost.

2. **Bills of quantities** 工程量清单，工程量表

   A bill of quantities is a form in the tender document carrying the items of the scope of work to be completed by the contractor for the bidder to prepare a bid. Usually this form includes the description and estimated quantities of each item of work, and there are blank space of the unit price and total price following each item to be filled up by the bidder. This form is an important data on which the employer's consequent evaluation of the tender will be based.

3.3 Management contracting

The characteristics of a management contract are that the client engages the management contractor to participate in the project at an early stage, contribute construction expertise to the design and manage the construction. Because of these requirements, it is normal for the management contractor to be an experienced builder or construction company, but this is not a pre-requisite. The management contractor is not employed for the purposes of undertaking any of the works, but solely for managing the process. In fact, management contracting is a procurement method consisting of 100% sub-contracting. Every item of building work is sub-contracted to works contractors. The role of a management contracting is like that of a professional or consultant, so it can become the cheapest procurement method by which a lot of labor force, plant, and equipment could be saved, and meanwhile the operating costs reduced, accordingly the management contractor's risk associated with the project is also reduced because the contractual risk in connection with the construction of the building is distributed entirely between the client and the works contractors, i. e. the management contractor is not legally responsible for the defaults of works contractors. This leaves the management contractor with very little contractual risk.

Some projects with high complexity and risk may lead to an inflated price, as the tendering contractors try to cover the risks, but the level of uncertainty associated with contractual risk is reduced for an experienced client who has large resources and is in a stronger position to bear risks. In this case, a management contracting may be the first choice.

As for the price, the management contracting may let the tendering works contractors quote the lowest price because each package of work will be undertaken by a specialist who will be more competitive for that part of the works. Had that same works contractor tendered for the whole of the works, then they would not have been as competitive.

While the management contractor has an obligation to control costs, the employer has to pay whatever the management contractor spends, plus an amount for the management fee, which may be either a lump sum or a percentage of the prime cost. This means that the contract between the employer and the management contractor is a cost reimbursement contract.

Apart from questions of price, if a works contractor claims from the management contractor as a result of the default of another works contractor, the management contractor is committed to pursuing recovery from the defaulting works contractor. If this is not possible, for example because the works contractor is insolvent, then the employer has to make up the shortfall. This is, of course, a vital transfer of risk from the management contractor to the employer.

By using this form of procurement, the employer may at any time determine the employment of the management contractor under the contract. Where the employer exercise this option, the benefit of any outstanding contracts for works or supply is automatically assigned to the employer. This is subject only to the rights of the relevant works contractors to make reasonable objection to any further assignment by the employer.

Management contracting is designed to encourage overlap between the process of design and construction. For this reason, the chances are that a building will progress from conception to completion more quickly than under a traditional arrangement.

As for the quality of both materials and workmanship, the management contractor shall be fully liable to the employer for any breach of the terms of this contract, including any breach by any of the works contractors under the terms of their works contracts. However, this liability of the management contractor is limited whatever can be recovered in turn from the defaulting works contractor. The philosophy underlying this stipulation has been expressed by saying that no independent liability can be attached to the management contractor for any defects in workmanship and materials, on the basis that it is the works contractors who are the parties responsible for the completion of the works.

The following are the suitable circumstances in which to use management contracting, but not all these conditions are necessary for a successful management contract, it is perhaps only when the majority of these factors act in combination that the management contract should be used. The conditions mentioned are as follows:

1. the employer wishes the design to be carried out by an independent architect and

design team.

2. there is a need for early completion.

3. the project is fairly large.

4. the project requirements are complex.

5. the project entails, or might entail, changing the employer's requirements during building period.

6. the employer requiring early completion wants the maximum possible competition in respect of the price for the building works.

Words, Phrases and Expressions

expertise *n.* 专门知识和技能（professional knowledge and skill）
pre-requisite *n.* 前提，先决条件（precondition）
works contractor 工作承包商（working contractor）
operating cost 运营成本，工作成本（working cost）
contractual risk 合同风险（contract risk）
prime cost 最初成本（initial cost）
pursue *vi.* 追索（to get back，get back）
insolvent *adj.* 无力偿付债务的，破产的（bankrupt）
shortfall *n.* 差额（balance，margin）
outstanding *adj.* 未履行的，未偿付的（not yet settled or completed）
assign *vi.* 转让，交给（to transfer）
progress *vi.* 进展，前进（to proceed）
conception *n.* 概念，构想（forming an idea）
combination *n.* 结合（integration）
entail *vi.* 使…．成为必要，需要（to necessitate）

疑 难 词 句

1. Had that same works contractor tendered for the whole of the works, then they would not have been as competitive.

译文：如果这家相同的工作承包商对整个工程投标，他们就不会具有如此的竞争力。

Notes：Those contractors are specialist contractors, meaning that they are specialized and experienced in their own fields but not so good at all the respects involved in a project. That is why they cannot become a general contractor for a project.

2. This is subject only to the rights of the relevant works contractors to make reasonable objection to any further assignment by the employer.

译文：但相关的工作承包商有权合理地拒绝业主对该未履行完毕的合同的利益的再次转让。

Notes：This sentence means that after the employer is assigned the outstanding contracts under which the works contractors are the real or actual contractors, the employer has no right to further transfer these contracts to the third party (to ask any other contractors to complete these outstanding contracts) without prior consent of the relevant works contractors.

3. However, this liability of the management contractor is limited whatever can be recovered in turn from the defaulting works contractor. The philosophy underlying this stipulation has been expressed by saying that no independent liability can be attached to the management contractor for any defects in workmanship and materials, on the basis that it is the works contractors who are the parties responsible for the completion of the works.

译文：但是，管理承包商的责任仅限于他能从违约的实际承包商那里追索回来的赔偿额度。该规定的理论基础是：管理承包商不对工艺和材料的任何瑕疵单独负责，因为实际的承包商才是履行该合同的责任方。

4. The employer requiring early completion wants the maximum possible competition in respect of the price for the building works.

译文：业主既要求早日竣工又希望在价格方面尽可能地得到充分的竞争。

Notes:

1. Management contractor

   The job of a management contractor is only to give advice to the design and manage the construction of a project, he does not undertake any of the works but only manages the process. All the physical works are subcontracted to the works contractors who actually do the works of a project.

2. Works contractor

   Works contractors are sometimes called doing contractors who actually fulfil the exact construction works of a project and who are actual contractors of a project. But under the management contracting arrangement, the employer signs with the management contractor and the management contractor signs with all the works contractors. Therefore the works contractors are like sub-contractors in the traditional general contracting.

3. Cost reimbursement contract

   This type of contract is also called a cost-plus contract. Under this arrangement, the employer pays the contractor for costs necessarily incurred in execution of the contract and either a fixed fee or a fee based on a percentage of the cost of the contract.

3.4 Construction Management

As construction projects became more complex and design and construction professionals became more specialized, there was a strong need for a professional on a project who could coordinate all other participants such as owners, architects, and contractors to complete the project successfully. Otherwise, the separate prime contractors who are assigned their own tasks could not deal with each other very well because their responsibilities for the project are independent from those of other contractors and they usually could not interact with other contractors very well.

As construction projects continue to increase in complexity, the demand for managers to oversee these projects will also increase. Besides, sophisticated technology and the proliferation of laws setting standards for buildings and construction materials, worker safety, energy efficiency, and environmental protection have further complicated the construction process, which would also increase the need for construction management.

Under CM system, the fast path method is used in construction, a CM firm with construction experience is employed at very beginning so that they can participate in the designing work and offer some construction suggestions for the designer to consider while designing the works, and then the CM firm is in charge of management of all construction process. This system changes the traditional tender method where the tender process starts after design is completed into the contracting by stages where a group is formed jointly by the owner, CM firm, and designer, which is jointly responsible for the organization and management of the layout, designing, and construction of the works. Under this system, CM firm is in charge of overseeing, coordination, and management of the works. During the construction, CM firm meets with the contractors regularly, supervises costs, quality, and schedule, and predicts and monitors the change in cost and schedule.

The advantage of this system is shortening of the period from layout and design to completion of the works, thus saving the construction cost, reducing the investment risk, and benefiting early.

Construction management is conducted by construction managers, whose role is to coordinate and supervise the construction process from the conceptual development stage through final construction, making sure that the project gets done on time and within budget. They often work with owners, engineers, architects, and others who are involved in the construction process. Given the designs for buildings, roads, bridges, or other projects, construction managers oversee the planning, scheduling, and implementation of the project to execute those designs.

The single most important distinguishing feature of construction management and the one which distinguishes it most clearly from management contracting is that the employer places a direct contract with each of the specialist and trade contractors.

The construction management method of procurement is most suitable where some or all of the following circumstances are present:

1. the employer is familiar with construction, and knows some or all of the professional team.

2. the risks associated with the project are dominated by timeliness and cost (e. g. the employer may be a private-sector employer requiring a commercial building).

3. the project is technologically complex, involving diverse technologies and subsystems.

4. the employer needs to retain the right to make minor variations to requirements as the project proceeds.

5. the nature of the project is such that it makes sense to separate professional responsibility for the design of the project from professional responsibility for the management of the project.

6. the employer requires an early start on site.

7. the cost to the employer needs to be competitive, but the control of cost in terms of securing 'value for money' is more important than simply securing the least possible cost.

These criteria are somewhat similar to those favouring the use of management contracts. As with management contracting, some or all of these circumstances may combine to make construction management the best option.

If construction management is to work properly, the employer must take an active role in the management of the process. And therefore, the regular and effective feedback is needed from the project team to the employer, and the employer may appoint his own project manager to help managing the process. So this is not a procurement method for the inexperienced employer.

For construction management to succeed, the entire team, including the employer, should understand that the process is intended to be speedier than other procurement methods and the need for speed may compromise decisions about least cost. Fast projects need quick decisions. This precludes the investigation of problems from every angle.

The construction management is suited to both technically complex projects and the technically simple projects, because this management structure is adaptable throughout the life of the project and that the significance of each party's input will vary from one stage to another.

This process accommodates minor variations while the project proceeds.

Not all projects are suited to construction management. The situation most suited to construction management is where design is neither an intrinsic part of the project nor the overriding feature of the job and almost all the activities related to design have been delegated to consultants, so that the problem of co-ordination is not a design problem but a pure management problem.

In many cases, 'value for money' is more important than least cost. Construction management has the capacity to allow for contracts either to be put out to bid or negotiated. Selection of trade contractors can be based on the most appropriate criteria for each package.

In the case of construction management, the speedy progress is desired by the employer, which requires the skill and experience of the construction manager to work out a proper programme which cannot be so tight as to result in inflated tenders from the trade contractors, and too loose as to result in a slow project.

By management construction, the direct contract between the employer and the trade contractor, without the intervention of a main contractor, has some advantages in terms of money as below:

1. prompt payment of certificates is ensured;

2. performance of the trade contractor is improved;

3. costs of finance to the trade contractor is minimized;

4. high degree of confidence in each of the contract sums, because the contract sum in each case cannot be altered except by express provisions of the contract;

5. the final price of the project should be predictable provided that the employer restricts the use of variations.

3.5 EPC

It is an Engineering-Procurement-Construction system. Under this system, Engineering not only involves with the designing work, it also probably includes the general layout of the work scope of the works and planning and concrete work of the organizational management of the overall construction works. Under EPC system, the owner only describes generally the investment purpose and requirements, all the rest work will be completed by EPC contractor; the owner does not need to employ the supervising engineer to manage the works, but may appoint a owner representative or by himself to manage the works; the contractor bears the most risks such as design risk, elemental risk, unforeseeable difficulties, and etc. ; and the lump sum contract is normally adopted. Under the traditional delivery system, the materials and plant are usually purchased by the general contractor of the project, but meanwhile the owner can withhold the purchase right of some important plant and special materials as well as the risk associated therewith during the execution of the project. Under EPC model contract, however, the contractor should be responsible for all the design work and bear all the responsibilities of the works, the owner therefore should not interfere with the contractor's job. The principle of EPC system is that the owner participates little in management of the project for the contractor has assumed most risks. The owner's job is mainly to accept the project after it is completed.

Words, Phrases and Expressions

fast path method 快速路径法
distinguish *vi.* 区别（to differentiate）
feedback *n.* 反馈信息（information，opinion）
compromise *vi.* 折中，削弱（to weaken）
preclude *vi.* 阻止，妨碍（to prevent，to disturb）
adaptable *adj.* 容易适应的，可调整的（adaptive，adjustable）
accommodate *vi.* 容纳，接受，允许（to allow，accept）
value for money 合算，值得，性价比高（worthwhile，cost-effective）
inflated *adj.* 高价的，涨价的（high-priced）
predictable *adj.* 可预测的（expected）
elemental *adj.* 自然力的（physical）
model contract 标准合同，格式合同（standard form contract）

疑 难 词 句

1. The nature of the project is such that it makes sense to separate professional responsibility for the design of the project from professional responsibility for the management of the project.

译文：该项目的性质导致了项目中设计的专业责任与管理的专业责任是分开的。

Notes：The nature or characteristics of a project can determine whether the professional responsibility for the design of the project is separated from that for the management of the project. The construction management is suited to the project where the professional responsibility for design is separated from the professional responsibility for the management.

2. The cost to the employer needs to be competitive，but the control of cost in terms of securing 'value for money' is more important than simply securing the least possible cost.

译文：对业主来讲，成本当然需要具有竞争力，但是用'资金价值'来控制成本比简单地获得尽可能低的成本更重要。

Notes：'value for money' means cost-effective，getting higher value with a certain amount of money or getting a certain value with less amount of money.

3. These criteria are somewhat similar to those favouring the use of management contracts. As with management contracting，some or all of these circumstances may combine to make construction management the best option.

译文：这些标准与前述的使用管理承包法的适用标准有些相似。如同管理承包法一样，只有当这些情况中的一些或全部都存在时，施工管理法才是最好的选择。

4. The situation most suited to construction management is where design is neither an intrinsic part of the project nor the overriding feature of the job and almost all the activities related to design have been delegated to consultants, so that the problem of co-ordination is not a design problem but a pure management problem.

译文：最适合施工管理法的情况是：当设计既不是项目的本质部分又不是项目的最重要的特点，并且几乎所有与设计有关的活动都委托给了咨询顾问。因此，协调问题不是一个设计问题而是一个纯管理的问题。

Chapter 4 Tendering and Contract Formation

4.1 Agreement

A contract is a legally enforceable agreement. An agreement is usually defined in terms of an 'offer' made by one party and an 'acceptance' of that offer by the other.

4.1.1 Offer

The traditional distinction drawn by the law of contract is between an 'offer' and an 'invitation to tender'. Offer is quite different from an 'invitation to tender' in that an offer is turned into a contract immediately on its 'acceptance' by the person to who it is addressed, but an 'invitation to tender' has no such power - it is merely a stage in negotiation, inviting the other party to make an offer.

Letters of intent

In the construction industry, it is usual to find work on a construction project started before (even long before) a formal contract is signed or even after the job is finished. In such circumstances the employer may be asked to write a 'letter of intent', which indicates a firm intention to award the contract in question to that contractor. However, the following principles shall be observed when writing a letter of intent:

First, a letter of intent in itself does not usually give rise to any legal rights or obligations. This is because the letter, by stating that there will or may be a contract in the future, is clearly indicating that there is no such contract at present. A contractor who carries out work on the basis of such a letter will be entitled, under a legal doctrine called restitution, to be paid the reasonable value of the work carried out, but there are no further legal consequences.

Estimates and quotations

It is sometimes said that, in the context of construction, the law draws a sharp distinction between an 'estimate', which is a mere invitation to treat, and a 'quotation', which is an offer in the legal sense. While there may be at least a grain of truth in this, in that a court will tend to assume that this is what the parties intend, the labels cannot be regarded as conclusive. This point is clearly illustrated by many old cases in the past. For example, a defendant, in response to an architect's invitation to tender, wrote a letter headed 'Estimate' which stated: 'Our estimate to carry out the works according to the drawings and specifications amounts to $ 1230. The plaintiff employer replied 'accepting' this figure, but the defendant thereupon purported to withdraw their 'estimate'. Then, when the plaintiff sued for the extra cost involved in having the work done by another contractor, it was held that the defendant was liable. Notwithstanding its heading, their letter was an offer.

Where an estimate is not treated as an offer, it is obvious that the employer cannot, by purporting to 'accept' it, ask the contractor to do the work at the price stated, or indeed to do any work at all. Further, even if the contractor does begin work after giving an estimate, neither party can insist that payment be measured by the amount of that estimate. If there is no further agreement, the employer will be liable to pay what the court regards as a 'reasonable sum'. This may be higher or lower than the quoted figure.

4.1.2 Acceptance

Assuming that an offer in the full legal sense has been made, a binding contract will come into existence when this is accepted by the other party. However, it is important to note the requirements for a valid 'acceptance' in this context.

Acceptance must be certain and unambiguous

In order to create a contract, a party's acceptance must unequivocally relate to the other party's offer. Further, the resulting agreement must be certain in all its essential terms. If these requirements are not met, there will be no contract.

Acceptance must be ***unconditional***

Where what looks like an acceptance (or, for that matter, an offer) is made conditional upon the happening of some event, no contract is created at that point. For example, the use of the phrase 'subject to contract' in negotiations will in normal circumstances prevent the document in which it is contained from being treated as a binding offer or acceptance. However, it is perfectly possible for parties to make an informal agreement at the beginning and for this to be legally binding, even though they expressly know that their agreement will be put into formal shape at a later stage.

Apart from communications which are explicitly made conditional, difficulties of distinction may arise in the case of negotiations which last a long period of time. Apurported acceptance which does not exactly refer to the offer will take effect as a counter-offer, destroying the original offer and leading to a contract only if it is itself accepted by the other party. When faced with a long series of letters, each of which accepts some of the other party's terms and proposes alterations to others, a court must try to see whether, at any given points, the parties are in complete agreement on everything which is at that time regarded as an essential term. If this is so, then a contract may be regarded to have been formed, and the effect of later negotiations can be neglected.

Acceptance by conduct

One important principle is that the contractual problems arising out of a lengthy exchange of correspondence may sometimes be resolved by examining the conduct of the parties. It sometimes happens that the conduct of one party can show an acceptance of the terms as they exist at that time. This principle is of great importance in construction cases. Work is frequently started before a formal contract is drawn up and signed. If a dispute arises before this happens, a court may well be able to find sufficient 'acceptance' of terms contained in earlier documents, either by the contractor in starting work or by the employer in giving possession of the site.

Therefore, cognizance of an acceptance is not based on its name or expression but on the conduct or behavior by the parties involved therein. Sometimes, a purported acceptance is not considered to be a real acceptance by the court, yet a statement or a correspondence, though they are not called an acceptance, would be taken as an acceptance by the court just because the judge thinks that the document in question or any other information can prove that the both parties thereto have previously admitted it as an acceptance.

Words, Phrases and Expressions

enforceable *adj.* 可强行的（compelled, compulsive）
offer *vi.* 出价，报价（to quote）
invitation to tender 招标（inviting others to make an offer）
acceptance *n.* 接受，认可（agreeing to, confirming）
entitle *vi.* 使有资格，使有权（to give right to）
restitution *n.* 补偿（compensation）
purport *vi.* 声称（to profess）
binding *adj.* 有约束力的（involving legal obligation）
valid *adj.* 有效的（effective）
unequivocal *adj.* 不含糊的（unambiguous）
essential *adj.* 实质的，本质的（intrinsical, substantial）
unconditional *adj.* 无条件的（termless）
explicitly *adv.* 明白的，明确的（clearly, specifically）
alteration *n.* 更改（change）
correspondence *n.* 信函，通信（letter）
conduct *n.* 行为（behavior）
cognizance *n.* 审理，认定（affirmation）

疑 难 词 句

1. While there may be at least a grain of truth in this, in that a court will tend to as-

sume that this is what the parties intend, the labels cannot be regarded as conclusive.

译文：当其中可能至少有一点实情时，这时法院都会倾向于认为这点实情就是双方的真实意愿，而称号则不能作为判断事实的依据。

Notes：Where a court is judging a case, the judge will reason from the context of the essential truth in a comprehensive way instead of from the labels or headings and so on. Therefore, sometimes either a so called quotation may be regarded as an estimate or a purported estimate may be deemed a quotation in a court.

2. A purported acceptance which does not exactly refer to the offer will take effect as a counter-offer, destroying the original offer and leading to a contract only if it is itself accepted by the other party.

译文：如果同意接受的内容并未完全与报盘的内容一一对应的话，这种接受只能作为一个还盘来看待，它推翻了原有的报盘并将导致只有当对方接受后才能成立的另外一个合同。

Notes：An acceptance, though it does not purport to attach any condition, which is not in complete agreement of all the terms or essential terms of an offer will be deemed a conditional acceptance. That is to say, such an acceptance is not a real acceptance, because an acceptance must be unconditional.

3. One important principle is that the contractual problems arising out of a lengthy exchange of correspondence may sometimes be resolved by examining the conduct of the parties.

译文：一个重要的原则是：由大量的信函交流产生的合同问题有时可能会通过审查各方的行为来得到解决。

Notes：An established principle is that a judgment on whether an acceptance is constituted is not based on what the involved parties have said but on what they have behaved.

4. If a dispute arises before this happens, a court may well be able to find sufficient 'acceptance' of terms contained in earlier documents, either by the contractor in starting work or by the employer in giving possession of the site.

译文：如果纠纷发生在合同签署前，法院完全能够在以前的文件中找到足够的有关“接受”的条款，这些文件既有可能是承包商开工时发出的，也有可能是业主在交出工地时发出的。

Notes：This sentence means that although a formal contract may not be drawn up before a dispute arises, the court still can make a judgment as if the formal contract had existed provided that the court is able to find the sufficient evidence of 'acceptance' in other documents previously exchanged between the parties.

4.2 Contracts made by tender

Most contracts for building projects are created by the process of tender, because tender enables the employer to choose the contractor who may best response to the requirements of the employer.

4.2.1 Purpose of tendering

There are two purposes of a tendering procedure. First, a suitable contractor should be selected at a proper time for the project. Second, the offer of a price is required from the relevant contractor at an appropriate time. This offer (tender) will be the basis for the subsequent contract.

The wise contractor will consider the conditions of contract when calculating the contract price. The way in which risks are allocated in a building contract will have a big impact on the contractor's pricing strategy. Along with the conditions of contract, the contract drawings and the bills of quantity (if used) constitute the contractual arrangements. The contractual arrangements are reflected by the procurement strategy of the employer. This is often not an explicit decision on the part of the employer, especially if the employer is unfamiliar with the construction industry. The contractual conditions give a legal basis to the rights, liabilities and duties of the parties to the contract. They form the basis of the organizational strategy adopted for the project. But there are very definite differences between contractual arrangements, procurement strategies and tendering procedures. It is important to remember that there is no direct relationship between the type of tendering procedure and the form of contractual arrangement. Further, there is often little relationship between the form of contract and the procurement strategy, even though there ought to be.

The tendering procedure, then, consists of two parts. First, the contractor has to be chosen, and second, the price for the contract works has to be estimated so that it can form a basis for the contract, although the two events are often simultaneous, they do not have to be. The tendering process represents the beginning of a contractual relationship. Therefore, the tender stage of a construction project is as much a beginning as an end. Too often, there is a tendency in construction for some professional consultants to regard the tendering procedure as the end of their involvement with the project.

4.2.2 Tendering procedures

A variety of tendering procedures have developed in the construction industry. The main character is the extent of competition. There is a very strong tradition that the best price can be gained by making contractors bid for work, so that the lowest price obtains the job. However, there is also an increasing level of dissatisfaction with competition, because all that the employer is sure to get is the lowest tender price. Usually, this is not the actual price that will be paid for the works once they are completed. For example, a contractor who bids too low, in order to be sure of getting the job, may later find that the job actually loses money. Contractors who bid as low as this tend to be those who are in bad need of the project because they are on the brink of insolvency. While a low tender may seem attractive to the employer at tender stage, it becomes very bad when the contractor abandons the works when he realizes that he would lose if he continues to work, and the employer has to employ others to complete the works. This can cost the employer a lot more than a higher initial bid would have cost. Some common tendering procedures are briefly described below:

Open tendering

This method was probably the 'conventional' method until more sophisticated ways were accepted. The process begins by the placing of an advertisement in the technical or business periodicals. The advertisement will carry brief details of the location, type, scale and scope of the proposed project. Contractors who are interested in bidding for the work can apply for the tender document. There will usually be a non-refundable charge for this documentation to prevent people from applying out of idle curiosity. Either the advertisement, or the documentation, will explicitly state that the employer has no obligation to accept the lowest tender, or indeed any tender.

Open tendering is a request for tenders from all the interested bidders. This approach can be unprofitable because it cannot ensure high quality works. According to statistics, with open tendering, only about one in twenty contractors' bids are successful. The preparation of such tenders brings the industry an unnecessary burden of time, effort, and expense. This expense ultimately is passed back to the clients of the industry. Because of the indiscriminate nature of open tendering, contractors can be awarded the project for which they are not properly competent, in terms of either resources or experience. Although an employer is not bound to accept the lowest bid, the accountability for the public expenditure is under a lot of pressure to accept the lowest. When the lowest bid is accepted, this can easily result in the employer awarding the contract to the contractor who has the least knowledge of the complexities of the projects; or the greatest willingness to take risks; or the lowest but unrealistic bid price of all the bidders. And this unrealistic bid price would easily lead to the contractor's lose money or finish a low quality project. It would be very difficult for the employer to benefit from these factors and almost impossible to get the result of the best value for money.

Because of the problems associated with open tendering, its use has been declining in recent years. Indeed, as an alternative, the selective rather than indiscriminate tendering has been increasingly used nowadays. However, the open tendering, for its unique advantages, is still occasionally used to obtain tenders for building work. And further, the indications are that its use may have increased for public sector work because of its impartial and economical advantages. The intention is that open tendering should become the main form of public procurement, even though some use of restricted or negotiated procedures is permitted.

It is unfortunate that having recognized the large number of problems associated with open tendering, there are still similar pressures to apply this approach in the international market, especially where organizations such as the World Bank are involved in many developing countries, as the projects sponsored by the World Bank should all be procured by means of open tendering according to the Procurement Guideline of the World Bank.

Single stage selective tendering

The first step to resolve the problems of open tendering is to restrict the number of tenderers being invited to submit bids. This is the purpose of single stage selective tendering, which is also known simply as selective tendering. Basically, it consists of preselecting a limited number of contractors to tender for the work. An employer who builds regularly will usually have an approved list of contractors, from which a short list can be drawn up. The local government will usually set up specific procedures for a contractor to be entered on the approved list.

This approved list will consist of contractors of established skill, integrity, responsibility and proven competence for work of the features and size contemplated and there should be a short list. The code of procedure specifies which should be considered when considering a contractor for inclusion on the list. Clearly, once the pre-qualification process has been done properly, any of the contractors is satisfactory to the employer. Therefore, tenders may be considered on price alone, and the lowest one may be selected. Provided that tenders are checked carefully for mistakes by the bidders, and that the list of approved tenderers is regularly revised, some of the worst problems associated with inappropriate selection can be avoided.

However, other problems may still take place. Some of the major problems on construction projects occur because the contractor does not join at the design stage so that the design team cannot benefit from the contractor's knowledge and experience at a stage of design when it would be most useful. And issues such as continuity of work, production cost savings, 'buildability' and sub-letting specialist items can have a significant impact on the final price of a project. The successful management of some types of construction project can be totally dependent on getting this kind of knowledge into the design team at an early stage in the process. Now, a variety of procurement processes have emerged to deal with this, aiming to separate the processes involved with selection of the contractor from the processes for determining the pricing mechanisms to be used when paying the contractor.

Two stage selective tendering

One of the techniques used to involve a contractor at an early stage is known as two stage selective tendering, sometimes called 'negotiated tendering'. It must be stressed that the purpose of this system is not to involve the contractor with responsibility for design. It is to get the main contractor involved, in an advisory capacity, before the scheme has been fully designed. This may lead to complaints from contractors, or more usually, specialty sub-contractors, that they are brought in early to help with design without being paid for design services.

In this method, the process is divided into two stages. The first stage is a process for the selection of a contractor, and for the establishment of a level of pricing for the following negotiation. It is at this stage that the competitive selection takes place, based upon a minimum amount of information which indicates simply the basis of the layout and design of the works. The detail needs only to be enough to provide a basis for competition, and a basis for the negotiation of the price during the second stage. In the second stage the pre-contract process is completed. The employer's professional team collaborates with the selected contractor in the design and preparation of construction drawings for the whole project. Also bills of quantities for the works are developed and priced upon the basis of the first stage tender. The result of the second stage is an acceptable sum for inclusion in a form of contract and complete contract documentation is worked out with the contractor.

This kind of tendering is used where the building works are of a very complicated nature; where the magnitude of the work may be unknown at the time for selection of the contractor; or where an early completion date is very important.

Negotiation

A more effective approach to the selection of the contractor is offered via negotiation. This approach is more suited to procurement strategies which are markedly different from traditional methods. Construction Management, for example, involves the use of direct contracts with different specialty contractors. Management contracting involves the early appointment of the main contractor whose responsibilities are not the same as other more traditional forms of contract. These newer forms of procurement demand a less adversarial approach at all levels and all stages in the building process. The inherent flexibility demanded by such approaches means that there is no standard method for negotiating a contract.

A variant of negotiation occurs with serial contracting, where contractors are asked to bid for a project (perhaps in competition) on the basis that, if they build this one satisfactorily, others of a similar type will follow and the same bill rates will be used. In the light of some of the recommendations in some technical periodicals about partnering, and also in the light of the experience of many sophisticated clients of the industry, this approach brings many benefits. Not least among them is that the desire to preserve continuing business relationships tends to be stronger than the need to fight particular claims or disputes. Thus, the parties to the contract are often focused on co-operation rather than conflict. Another advantage of serial contracting is that contractors have an opportunity to learn more about the client organization and there are opportunities to repeat details from on project to another. This can potentially enable significant cost savings brought about through repetition of tasks.

Experience has shown that negotiation is one of the most effective ways of selecting a contractor for 'non-traditional' procurement approaches. In these cases, the deal is negotiated as the relationship develops. It seems that the single most important factor of such a relationship between the employer and the contractor is familiarity and friendship. They have worked together before, and they expect to work together again in the future. As with serial contracting, the preservation of a long-term business relationship is more important than simply getting the lowest price for the employer, or the highest profit for the contractor. Because of this, it is essential that the employer is familiar with some or all of the building process, the professional team, the contractor and the specialist sub-contractors.

Joint ventures

Certain construction projects are so complex that it is difficult to distribute liability between the consultants. One solution is to give the whole responsibility of the design and construction of the project to a joint venture between all the consultants and some or all of the specialist contractors. This is more like the 'management-based' methods than so-called 'traditional' methods. It involves all of the parties to the Joint Venture agreement bear joint and several liability for the design and/or implementation of the project. It is not necessary to set up a special joint venture company to do this, but it is possible. The agreements between the parties to the joint venture will have to be very carefully examined, and probably backed up by performance bonds and/or parent company guarantees.

Words, Phrases and Expressions

condition *n.* 条款，条件（term, provision）
strategy *n.* 策略（policy）
procurement *n.* 采购，获得（purchase, acquirement）
simultaneous *adj.* 同时的（coinstantaneous）
competition *n.* 竞争（rivalship, contest）
abandon *vt.* 放弃（to give up）
initial *adj.* 最初的（original）
conventional *adj.* 常规的，传统的（traditional）
sophisticated *adj.* 高级的，尖端的（advanced）
advertisement *n.* 广告（ad）
scale *n.* 规模（size）

non-refundable *adj.* 不可归还的（non-returnable）
unprofitable *adj.* 无利可图的（unrewarded）
indiscriminate *adj.* 不分好坏的，不加区别的（making no distinctions）
accountability *n.* 有责任感（being state of responsibility）
unrealistic *adj.* 不实际的，不实在的（impractical）
selective *adj.* 选择的（chosen）
impartial *adj.* 公正的（fair）
negotiated procedure 议标程序（a tender procedure by negotiation between the employer and the selective potential contractor）
pre-selecting *adj.* 预选的（primary selecting）
approved list 经批准（审核）的名单
short list 短名单（人数已缩减的候选人名单）
established *adj.* 已制定的（set-up）
integrity *n.* 正直（honesty，rightness）
proven *adj.* 经验证或证实的（proved）
code *n.* 法规，章程（statute，constitution）
pre-qualification *n.* 资格预审（primary selection）
collaborate *vi.* 合作（to cooperate）
magnitude *n.* 规模（size）
flexibility *n.* 灵活性，可变性（changeability）
variant *n.* 变体（derived form）
serial contracting 连续承包
bill rate 费率
distribute *vt.* 分配（to allocate）

疑 难 词 句

1. Usually，this is not the actual price that will be paid for the works once they are completed.

译文：通常，这个投标价并不是工程竣工时业主应支付的实际价格。

Notes：In fact，the final actual price the employer should pay for the works is quite different from the tender price because the actual price will include many additional payments for the claims lodged by the contractor against a lot of unexpected events，such as variations by the employer，breach of contract by the employer，and other events which may occur during the execution of the project for which the contract entitles the contractor to the additional payments.

2. The intention is that open tendering should become the main form of public procurement，even though some use of restricted or negotiated procedures is permitted.

译文：其目的是：即使有时允许使用有限招标或议标，但公开招标应成为公共采购的主要形式。

Notes：'Public procurement' for this purpose means procurement by any association governed by public law (central or local government) and includes health authorities, education authorities and the police.

3. The successful management of some types of construction project can be totally dependent on getting this kind of knowledge into the design team at an early stage in the process.

译文：对某些类型的建设项目的成功管理可以完全取决于设计团队在项目的早期就掌握这种关于施工的知识。

Notes：The design team with the knowledge of construction can work out a more economical and more workable scheme which also will bring a minimum difficulties to the builder in converting the design concept to a real building.

4. Now, a variety of procurement processes have emerged to deal with this, aiming to separate the processes involved with selection of the contractor from the processes for determining the pricing mechanisms to be used when paying the contractor.

译文：现在，针对这个问题出现了各种不同的采购方式，目的是要将选择承包商的过程与确定给承包商付款时使用的定价机制的过程分开。

5. It is to get the main contractor involved, in an advisory capacity, before the scheme has been fully designed.

译文：这种方式是要让总承包商在设计方案完全出台前以顾问的角色介入。

6. The detail needs only to be enough to provide a basis for competition, and a basis for the negotiation of the price during the second stage.

译文：所需的详细资料只要足以能够为竞争和第二阶段的价格谈判提供依据就行了。

Notes：Generally, at the stage of tendering, it is not necessary for the employer to provide full details of information on the design scheme to the potential contractor. At this stage, only a minimum amount of information which is enough for competition and pricing is needed.

7. Not least among them is that the desire to preserve continuing business relationships tends to be stronger than the need to fight particular claims or disputes.

译文：此外，特别需要指出的是：维持这种连续不断的业务关系的愿望倾向于比赢得一些具体的索赔或纠纷的需要更强烈。

Notes：Throughout the execution of a construction project, although many claims and disputes may occur very often and they need to be resolved, both the employer and the contractor tend to keep a good relationship with each other.

4.2.3 Legal analysis of tenders

The adoption of a suitable tendering procedure will no doubt help to reduce the potential waste of time, effort and money. Nevertheless, there remain certain difficulties in the legal analysis of tenders, and these are discussed below.

The parties' obligations

The employer's request for tenders is regarded as an invitation to treat, according to the conventional legal analysis, and the tenders themselves as offers. Further, as a general principle, the employer is under no legal obligation to accept the lowest (or, indeed, any) tender submitted. In practice, this protection of the employer's position is frequently (though unnecessarily) made explicit when tenders are invited, by a statement that the employer does not promise to accept the lowest tender.

The result of this contractual arrangement is that the costs of tendering (which may be considerable, especially where substantial design work is required) are to be borne by the contractor. Of course, these costs can be included in the tender price and thus recovered by the successful bidder, but the unsuccessful competitors must in normal circumstances bear their own costs, unless a promise to pay by the employer can be implied. Such a promise might be implied where the preliminary work exceeds what would normally be expected, or where the employer can make some profitable use of it.

Withdrawal of tender

The basic contractual position is that, since a contractor's tender is merely an offer, it may be of course revoked at any time before it has been accepted. If this happens in respect of a main contract, the client may well be disappointed (especially if the tender in question was the lowest!). However, the client is unlikely to suffer any great financial loss due to this withdrawal. By contrast, the withdrawal of a sub-contractor's tender may have a disastrous effect on the main contractor, because the main contractor, having obtained a written estimate of costs from the subcontractor or supplier, and, on the basis of this offer, calculates the main contractor's tender for certain works. And then the main contractor's tender may be accepted by the client, but the subcontractor then may announce that they are increasing their prices (which in effect means that they are withdrawing their original offer). Thus, in the absence of a binding contractual obligation, the subcontractor is quite entitled to this revoke, which of course will leave the main contractor with an uneconomic contract to fulfil.

To avoid such an unfortunate case, a practical solution would appear to lie in requiring contractors to supply a 'bid bond', that is, a promise by deed not to withdraw the bid, backed by a financial guarantee.

4.2.4 Problems in the constitution of bids

Conventional text book wisdom describes contractor's tendering procedures in a very logical and objective manner. For example, it is always said that in pricing a bill of quantities a contractor will ascertain such things as the price of materials, the cost of labour and plant, the availability of resources. This information is used to constitute rates appearing in the bills. So, for example, if a certain quantity of brickwork is described in the bills, the contractor will determine the cost of the raw materials (based upon the relevant quantity discounts), the rates of pay for the skilled and unskilled labour force, the number of hours actually needed to put together a square metre of brickwork, the number of hours needed to mix the mortar, an amount of money for supervision, plant and so on. These amounts are added together in such a way as to identify a rate for a square meter of brickwork. This rate can then be inserted into the bill and multiplied by the specific quantity required.

This process, known as estimating, takes place for every item in the bill. The first problem is that it takes little account of the way in which contractors' costs are actually incurred; it is based only on historical information which does not relate the location of the work to its cost. An easy example is that brickwork on the second floor is cheaper than brickwork on the 17th floor, because of the distance that the materials have to be moved, but this difference is rarely reflected in the built-up rates.

The second, and more important problem, is the 'story behind tendering'. What is not often reported is that, although contractors have very detailed information related to costs, and can work out very accurate rates for the bills, these are frequently not the rates which actually appear in the bills.

Contractors know about costs (what they pay for their resources); they know aboutprices (what they charge for their product); and they know about value (what the client is willing to pay for a building). The difference between cost, price and value is rarely researched or studied outside quantity surveyors' offices. It is this fact which accounts for the difference between the text book method of tendering, and the additional process which is not usually reported. Upon being invited to tender, a contractor will first of all make a decision about whether or not the job is wanted. A contractor will almost never decline to tender, for fear of not being asked to tender again in the future. There is a certain amount of stigma attached to a contractor who declines to tender. If the job is not wanted the contractor will submit an inflated tender, called a cover bid. While the contractor will know what the rates in the tender ought to be, these will all be altered to make the final figure clearly too high to be acceptable.

If the contractor wants the job, then a further decision is about the situation of the market and an assessment of what this type of building is selling for at the moment. This assessment is then modified by the level of risk associated with the project, particularly in terms of the contractual conditions put forward by the employer. If the contract is risky, a considerable premium should be added to the contractor's bid, so that the risks are covered. The calculation is, in theory at least, similar to the actuarial calculations made by insurers when examining and pricing risks.

Finally, the contractor's own cash flow has a significant impact on the tendering process. Having a detailed knowledge of costs and finance, the contractor can predict the monthly net cash flow in or out of the project. If the project takes place near the end of a tax year, the contractor may want to reduce the level of profit appearing on the balance sheets for the purpose of reducing tax liabilities. This can be done by artificially reducing the rates for work at the beginning of the project and adding a corresponding proportion on to the rates for work later on in the project. Alternatively, the contractor may need to get cash in quickly to meet liabilities, or to show a good return for shareholders' dividends. In this case the rates at the beginning of the project can be increased, with a corresponding decrease in the rates at the end. The former process, is known as back-loading the bill, the latter is known as front-loading the bill. Neither process makes any difference to the contract sum, although both can have a significant impact on the continued survival of the contractor's business.

This brief analysis shows that, notwithstanding the contractor's detailed knowledge of costs and prices, the skill of a good contractor is in adjusting the contract sum at a level which will maximize the chances of winning jobs that are wanted, while ensuring that profits are adequate. The bill rates are then revised to add up to the desired contract sum, and internally adjusted to regulate the cash flow according to the financial position of the contractor.

This shows how dangerous it can be to attach too much credence to individual bill rates when analysing tendering procedures. It also illustrates one of the major pitfalls in using bill rates for valuing variations.

Words, Phrases and Expressions

substantial *adj*. 大量的（considerable, great deal of）
recover *vi*. 收回（to return, to get back）
revoke *vt*. 撤销（to cancel, to withdraw）
withdrawal *n*. 撤回，收回（withdrawing）
disastrous *adj*. 灾难性的（destroying）
bid bond 投标担保（bid security）
deed *n*. 契约（contract）
financial guarantee 金融担保（a guarantee by depositing a sum of money）
ascertain *vt*. 弄清，搞明白（to get to know clearly, to be aware of）
availability *n*. 可获得性（being available）
constitute *vt*. 构成（to form, to arrive at）
supervision *n*. 监督（supervising, overseeing）
estimate *vt*. 估算（to evaluate, to appraise）
built-up rate 组合费率（comprehensive rate, all-round rate, overall rate）
story behind tendering 投标背后的故事，投标潜规则（myth of tendering, unspoken

rules)
decline *vt.* 谢绝（to refuse）
stigma *n.* 耻辱（shame）
inflated tender 高价标（over-estimated tender）
cover bid 虚标（a false bid to hide something）
contractual condition 合同条件（条款）
put forward 提出（to propose）
balance sheet 资产负债表
artificially *adv.* 人工地，人为地（factitiously）
alternatively *adv.* 作为选择，另外一种选择（otherwise）
back-loading *adj.* 向后倾斜的，前轻后重的
front-loading *adj.* 向前倾斜的，前重后轻的

疑 难 词 句

1. The employer's request for tenders is regarded as an invitation to treat, according to the conventional legal analysis, and the tenders themselves as offers.

译文：按照传统的法律分析，业主的招标被当作邀请投标，投标本身被当作报盘。

2. Such a promise might be implied where the preliminary work exceeds what would normally be expected, or where the employer can make some profitable use of it.

译文：只有在前期工作超出正常的预想或业主对该前期工作可以有利可图（可以有利润）时，这种承诺才可以被认为是暗指（不言而喻）的。

Notes: Generally, as a widely acceptable practice, the tendering costs cannot be recovered by unsuccessful tenderers, except for very exceptional cases.

3. Thus, in the absence of a binding contractual obligation, the subcontractor is quite entitled to this revoke, which of course will leave the main contractor with an uneconomic contract to fulfil.

译文：这样，在没有合同义务的情况下，分包商完全有权撤销其报价，这当然会使总承包商不得不去履行一个经济上很不合算的合同。

4. The first problem is that it takes little account of the way in which contractors' costs are actually incurred.

译文：第一个问题是：这种方法很少考虑承包商的成本实际发生的环境。

Notes: The environment or situation where the costs are incurred is very important and will have a big impact on the calculation of the overall costs involved in any cost-incurring activity. For instance, the same activity of excavation may be conducted in surroundings of different rock types in different projects. Therefore, the overall costs for this activity should be adjusted accordingly.

5. What is not often reported is that, although contractors have very detailed information related to costs, and can work out very accurate rates for the bills, these are

frequently not the rates which actually appear in the bills.

译文：一个不太为人所知的情况是：虽然承包商都能掌握非常详细的成本信息，并且也能为工程量清单算出非常准确的费率，但这些费率常常并不是最终实际出现在工程量清单上的费率。

Notes：As a well-established practice in tendering for a construction project, all the rates calculated from the objective costs to be incurred during the execution of the project must be adjusted subjectively according to the tenderer's policy before they appear in the bills. The basis for this subjective revision are various, such as the desire to get the project, competition with other terderers, risks associated with the project, demand for cash flow, market situation, and etc.

6. The difference between cost, price and value is rarely researched or studied outside quantity surveyors' offices.

译文：在物料估算师的办公室以外，人们很少研究成本、价格和价值之间的差别。

Notes：Studying the difference between cost, price and value is not a easy question, because it not only lies in the definitions in textbooks, also it lies in the various complex factors in the social problems and economic problems associated with the different projects as well as the different relations between employers and contractors.

7. While the contractor will know what the rates in the tender ought to be, these will all be altered to make the final figure clearly too high to be acceptable.

译文：虽然承包商知道投标书中的各个费率应该是多少，但这些费率将会全部被修改，使得最终报出的价格非常高，以至于业主无法接受。

Notes：If a contractor wants to decline a tender invitation, submitting an inflated tender is one of the decent ways of refusal.

8. The calculation is, in theory at least, similar to the actuarial calculations made by insurers when examining and pricing risks.

译文：这种计算方法至少在理论上与保险公司在做风险考核和风险定价时所做的精算相类似。

9. Neither process makes any difference to the contract sum, although both can have a significant impact on the continued survival of the contractor's business.

译文：尽管这两种方法都对承包商业务的延续有重大影响，但都不会影响合同总价。

10. The bill rates are then revised to add up to the desired contract sum, and internally adjusted to regulate the cash flow according to the financial position of the contractor.

译文：然后，修改工程量清单的各项费率使之加起来达到所期望的合同总价，再根据承包商的财务状况进行内部各项之间的费率调整来调节现金流。

Notes：The front-loading and back-loading of the bill rates at the tendering stage is an effective way to regulate the cash flow of the project to meet the demand for the contractor's cash flow control to solve the financial problems which might occur during the project execution. This way is only an internal solution which brings benefit to the contractor's inhouse management, not affecting the final tender price and therefore it has nothing to do with the success of winning the tender.

Chapter 5 Contract Terms

After a contract has been properly made, it becomes necessary to identify precisely what obligations it imposes on the parties - that is to say, what terms it includes. In considering this question, what emerges is that there are three types of contractual term: those express terms which are contained in the main contractual document itself; those (also express terms) contained in other documents to which the main contract document refers; and those terms which are implied by law.

5.1 Express Terms

5.1.1 Terms and representations

In many countries, the law does not as a general rule require a contract to be made in any particular form. Thus it is, in principle, quite possible for the parties' agreement to be spread over a number of documents. Even where the parties have eventually signed a document described as 'the contract', it is open to a court to decide that this document does not represent a complete contract, and that the true contract consists of that document plus other terms. However, a court will not be quick to do this, for there is a presumption that a written document which appears to contain all the terms of a contract in fact really does so.

The consequence of this view of contracts is to create a degree of uncertainty in respect of the status of a statement or promise which is made by one party to the other during the course of negotiations leading to the signing of a formal contract. Such a statement (which might be oral, or might equally be contained in, say, a letter) can be treated in law as a term of the contract. In this case, if it is untrue, the innocent party may lodge a claim for breach of contract by the other party.

Alternatively, if it has operated as an inducement to the other party to enter into the contract, the statement may be treated as a misrepresentation. In this case the legal remedies available will depend on whether it was made fraudulently, negligently or innocently. This is not the place to discuss the remedies for misrepresentation, which will be found in any textbook on the law of contract. But it is true that they are generally less beneficial to the innocent party than a right to sue for breach of contract.

In seeking to decide whether a particular statement is a contract term or a mere representation, the courts make use a number of principles. Some of these may point in contrary directions. It is for example presumed that a statement made at a very early stage of negotiations is probably a representation, while a statement made by an 'expert' is more likely to be treated as a term. However, what is very often decisive in the context of construction contracts is the strong presumption that, where the parties have signed what purports to be a complete contract document, this does indeed represent the whole of their contract. Therefore, any collateral statement or promise is likely to take effect as a mere representation.

Words, Phrases and Expressions

impose *vi.* 强加 (to force)
express term 明文条款 (a statement explicitly expressed, opposite to implied term)
representation *n.* 表述 (expression)
presumption *n.* 推测，认定
inducement *n.* 引诱
misrepresentation *n.* 误说，错误的表示或表达
legal remedies 法律补救
fraudulently *adv.* 欺骗地
negligently *adv.* 疏忽地
innocently *adv.* 无辜地
decisive *adj.* 决定性的

疑难词句

1. Even where the parties have eventually signed a document described as 'the contract', it is open to a court to decide that this document does not represent a complete contract, and that the true contract consists of that document plus other terms.

译文：即使各方最终签署了一个被称作“合同”的文件，法院仍然有权作出该文件并不能代表一个完整的合同的确定，并且还有权确定真正的合同是由该文件加上其他条款构成。

2. The consequence of this view of contracts is to create a degree of uncertainty in respect of the status of a statement or promise which is made by one party to the other during the course of negotiations leading to the signing of a formal contract.

译文：这种合同观点的结果是，给在为签署一项正式合同而进行的谈判过程中，一方向另一方作出的声明或承诺的地位造成了一定程度的不确定性。

Notes: Although some statement or promise made by any party during the contract negotiation are not included in the consequent formal contract signed by both parties, they may be regarded as an integral part of the contract if they can have an impact on the consequence of the dispute handling by a court.

3. It is true that they are generally less beneficial to the innocent party than a right to sue for breach of contract.

译文：通常来讲，它们（这些法律补救的办法）都不如提起违约诉讼对受害方有利。

4. Therefore, any collateral statement or promise is likely to take effect, if at all, as a mere representation.

译文：因此，任何相关的声明或承诺，如果有的话，都可能仅仅被当做一种表述（而不是合同）。

5.1.2 Contract documents

As mentioned above, a contract which is made in writing frequently consists of more than one document. Certain categories of document are routinely found in construction contracts and the main ones are described below.

Contract documentation is the means by which a designer's intentions are conveyed to the client, the statutory authorities, the quantity surveyor, the contractor and the sub-contractors. In addition, the other design consultants each add their own specialized information to the increasing body of documentation as the project progresses. From a management and administrative point of view, every item in this 'contract documentation' is important, because it describes and records all aspects of the project. However, all the items cannot come within the more limited legal definition of 'contract documents', a special term used in construction contracts to refer to those documents which define the contractors obligations. Whichever form of contract is used, the contractor's basic undertaking is to carry out the works in accordance with these contract documents.

As to what the term 'contract documents' includes, nearly all the standard form contracts identify the articles of agreement, the conditions of contract, the appendix, the drawings and the bills as key contract documents. However, there are different views between the forms of contract in whether or not such items as programmes, specifications and the like should be taken as contract documents.

The various elements of documentation will be examined briefly in isolation from each other, before looking at how they relate to each other and how the conditions of contract integrate them.

Articles of agreement

The articles of agreement record in general terms what the parties have agreed to do. They identify the parties to the contract, what is to be built (the contractor's obligation) and what is to be paid (the employer's obligation). They put these obligations into the conditions of contract and the other contract documents.

In defining the contractor's obligation, the articles do not descend to any great detail, but merely state, in a section starting with the word 'whereas', that 'the employer is desirous' of having the briefly described works carried out. Some standard forms also record that tenders have been accepted, that drawings have been prepared, and so on. This section is sometimes called the 'recitals'. The purpose of the recitals is to describe the background and the purpose of the contract. They help to provide a basis for interpretation of the detailed provisions contained in the contract clauses.

Conditions of contract

The conditions of contract are the very detailed clauses which materialize the articles of agreement. The purpose of the conditions is to amplify and explain the basic obligations which the parties undertake by signing the articles of agreement. The conditions also provide administrative mechanisms for ensuring that the correct procedures are observed. Effective contract clauses of this kind deal efficiently with what would otherwise be breaches of contract, and therefore ensure that the contract is kept 'alive'.

Appendix

Certain data and facts relating to the execution of a building contract will differ from one project to another. To enable standard form contracts to be used in spite of these differences, the facts concerned can be summarized in an Appendix to the contract.

The Appendix contains such items as the starting date (or date for possession), completion date (or duration of the contract), defects liability period and amount of liquidated damages, and etc. Some of the contractual conditions specify what is to do if the relevant entry in the Appendix is left blank. Some provisions are not applicable at all unless the Appendix is completed, such as the employer's right to delay possession of the site by up to six weeks.

The contents of the Appendix are of fundamental importance in assessing the amount and duration of responsibility accepted by the parties. It is therefore essential that close attention is paid to the entries in the Appendix.

Drawings

The preparation of drawings is generally, although not always, the responsibility of the design team. The technical complexity of most modern buildings makes it essential that this process (including all associated information, as well as the drawings themselves) is properly planned and organized. Detailed guidance on this is available from numerous sources (for example, FIDIC). As with all types of communication, clarity is essential in production of drawings. The aim is to transmit the information in a way that can be understood. Good communication will inspire confidence; bad communication will cast doubt on the ability and integrity of the designer.

Drawings have many functions and each function is usually fulfilled by a different drawing. First, they form a model of the designers' ideas and help to identify and predict problems with fabrication and with appearances. Scheme design drawings and 'artistic impression' drawings help to fulfil this function. Although there is little information on them which would enable a builder to erect the proposed building, they amplify and explain the basic nature of what will be required in the finished building.

Second, drawings are the vehicle by which the designers' intentions are conveyed to the contractor. The detail design drawings contain information which shows how the separate parts of the building interact with each other. The detailed information from specialist sub-contractors and from other designers is co-ordinated and presented through such drawings.

Third, drawings form a record of what has been done. These 'as-built' drawings are essential to the building owner as a basis for future maintenance of the facility, and may not be the same drawings that were used for the purposes of fabrication. They are also a useful control and check document to compare what has actually been done with what was originally planned.

The multiple uses which different drawings are required to perform, and the interaction between drawings and other types of documentation, have often been a source of problems. As a result, the Co-ordinating Committee for Project Information (CCPI) was formed. This committee has produced a set of guidelines for the production of drawings, specifications and bills of quantities. These currently represent best practice in the industry and should be essential reading for everyone concerned with the documentation of construction projects.

The designer usually retains copyright in the design and, on completion of the work, can insist that all drawings are returned. When this is the case, neither the contractor nor the client is entitled to use the information again, for example to construct an identical building, without first obtaining the permission of the designer.

Schedules

Schedules are tables of information which summarize the quantities and dimensions of certain kind of items, for example windows or ironmongery. Although they will rarely be cited as contract documents, they will often be observed and included with the sets of drawings, as they are invaluable in summarizing detailed information. Clearly, a drawing will show all the windows in a building, but it is very useful to extract all the details of sizes and numbers on to one table. Although dimensions can be summarized in tabular form, some information is best conveyed on a dimensioned sketch, such as the arrangement of opening in a window.

Schedules provide information to quantity surveyors, builders' buyers and others in a form which has advantages over drawings: mistakes are easier to check; getting estimates and placing orders for similar items is simplified; time-consuming searching through a specification is avoided. The aim of the schedule is merely to simplify the retrieval of information. It is not intended to take place of the bills of quantity, which will contain the full and final description of the work.

Specifications

Drawings provide information about the shape, appearance and location of the various components which have to be assembled but they convey little about the methods of construction, the quality of finishes and the workmanship to be employed. Assuming that such things are to be specifically controlled by the contract, rather than relying on the terms which would be implied at common law, this is a task for the specification. It defines the materials and products to be used, the standard of work required, any performance requirements and the conditions under which the work is to be executed.

Where bills of quantity are used for a project, information of this kind will be contained in the bills, and there will be no 'specification' in the form of a separate document. However, not all projects need full quantification by means of bills of quantity; many projects are quantified by drawings and specification alone. Where this is done, the specification may be required to play an even wider role. In order to prepare a useful and accurate specification, it is essential to be systematic and methodical. There are a variety of guidelines for a specification writer to observe.

Bills of quantity

The purpose of bills of quantity, and their status, may vary under different standard form contracts. However, when used, they are almost invariably specified as a contract document.

The contracts govern the preparation of the bills, typically specifying that bills have been prepared in accordance with the relevant standard method of measurement. Because the bills have such contractual significance, it is often necessary to be fully aware of what the relevant Standard Method of Measurement (SMM) contains. Any item of work that is not measured in line with the principles in the SMM must be expressed definitely in the bills, in terms of both the nature of the change as a result of not observing SMM and which items are affected by this change.

Bills of quantity originated for tendering purposes; indeed, engineering contracts still retain this emphasis. In many different forms of contracts, the bills of quantity have additional roles to play in the valuation of variations and interim certificates, and in the control of the works. In addition, various other categories of information are often learned from the bills, such as the location of work and basic cost planning data. Not surprisingly, bills of quantity are increasingly being seen as hopelessly inadequate for all of these conflicting needs. There are many ways of choosing a contractor, and many ways of fixing a price for the work to be performed. To use the bill as the sole vehicle for both these purposes is an over-simplification that can lead to many problems.

The bills typically consist of three sections namely preliminaries, preambles and measured works. These terms have specific meanings and it is useful to clarify their meaning:

• Preliminaries contain the definition of the scope of the works. In the bills, section A is given over to defining the preliminaries and general conditions. It includes, among other things, particulars of the project, lists of drawings, description of the site, scope of the work, details of documentation and management arrangements. The preliminaries often contain items for pricing which will be for general use for the whole project such as huts, fences and security.

• Preambles were once a separate description of the materials and workmanship to be employed in assembling the building and this was known as the 'specification' section of the bill. Therefore, it is likely that the term 'preambles' will be no longer used in building contracts.

• Measured works are the detailed quantification of the works. This section should be presented according to the rules stated in the contract.

Two further terms worthy of mention are prime cost and provisional sums. These terms, though very old, but are still commonly used and seems likely to continue:

• Prime cost (PC) sums are used for works under the building contract which are to be carried out under the direction of the main contractor by certain other persons, namely, nominated sub-contractors, nominated suppliers and so on. The nominated sub-contractors will usually have been selected prior to the selection of the main contractor. Since the contractor in this case does not price this work, an amount of money is simply included in the bills, to which the contractor can add a price for attendance (i. e. supervision, accommodation and plant) and profit. Prime cost is taken to mean that the contractor will be fully reimbursed for any valid expenditure.

• Provisional sums are used for work which has not been finalized or for costs which are unknown at the time the bills are prepared. They may simply be contingency sums, this amount of money is only a possible part of budget for the project, and the employer has not obligation to spend them. Spending these sums needs issuance of instructions by the contract administrator.

Certificates

There are several types of certificate in building contracts. They fall into two main categories: those which certify the quantity of work done to date, on which payment is calculated; and those which certify that an event has taken place. The former are called interim certificates, and their use is one of the most distinctive features of building contracts.

Interim certificates are used in the majority of building contracts, which enables the contractor not to finance the whole of the building work. The contractor will be paid for the amount of work done at regular periods; usually one month. When each payment is made, a small percentage of the money due to the contractor (called the retention) is held by the employer, and this is released when the work is completed. It is also possible for other amounts of money to be withheld, for delay or for defective work.

The second category of certificates consists of those which record an event (or non-event). They vary from one form of contract to another, but basically they are:

- Certificate of practical completion;
- Certificate of final completion;
- Certificate of non-completion.

Other documents

Usually, instructions contain much additional contract information, such information are more detailed than that in the tender documents. There are also other documents, such as schedules, further drawings and setting-out information. These may be necessary for the contractor to execute the work but such additional information must not impose extra work on the contractor. There will usually be an express term to this effect. If the contract administrator wishes to impose changes when issuing these instructions then a variation order must be issued to involve the changes.

There will typically be an obligation on the contractor to provide a programme of work. Although the contractor must provide a programme immediately after being awarded the contract, no extra obligations should be imposed on the part of contractor to do it before a certain legally binding date. The programme is essential as a management control means. It is also extremely useful to all participants in resolving claims for extra time. In order for it to be effective in these roles it must be amended as the project progresses and this is also usually a requirement.

5.1.3 Priority of documents

On a construction project of any substantial size, a great deal of detail which is found in the contract documentation gives great probability for discrepancies and inconsistencies. These may arise within one particular document, as where the contract bills contain two inconsistent provisions. They may also happen between two documents, as where something in the bills conflicts with something in the conditions.

The general solution of most standard form contracts to this problem is that in the event of finding a discrepancy, the latter must then issue instructions so as to resolve the discrepancy and, if these bring delay or disruption, the contractor will be entitled to an extension of time and to compensation for loss and expense the contractor should refer it to the contract administrator or the supervision Engineer under the FIDIC contract. Furthermore, if the instructions constitute a variation, the contractor will be paid at the appropriate rate for the work involved.

When required to resolve a discrepancy of this kind, the contract administrator should in general apply those principles of law which govern the interpretation of contracts. However, this is subject to the terms of the contract itself, and it is noticeable that these general principles are expressly modified by a number of standard form building contracts. For example, one of the common law rules is that written words prevail over typed words, and that typed words in turn prevail over printed words. That rule is based on the sensible idea that documents which are prepared for a specific job, rather than being taken 'off the shelf', are more likely to reflect the parties' true intentions. If it were to be applied to a building contract, it would mean for example that provisions in bills of quantity would override the printed conditions of contract.

This is emphatically not the case, at least where the main standard forms are concerned! The approach to this problem varies from one contract form to another. For example, some form of contract is fairly neutral, stating that all documents are to be taken as mutually explanatory, and another form provides categorically that the conditions shall prevail over all other documents. Similarly, JCT for instance, states that nothing in the contract bills shall override or modify what is contained in the articles, conditions or appendix. This wording does not prevent the bills from imposing extra obligations on the contractor. It does mean that where a particular matter is dealt with in the conditions, any special provisions on that subject in the bills are to be ignored.

However, some forms of contract designate their own priority of documents, for example, FIDIC conditions of contract specifies its priority of documents in accordance with the following sequence:

1. the Contract Agreement (if any);
2. the Letter of Acceptance;
3. the Letter of Tender;
4. the Particular Conditions;
5. the General Conditions;
6. the Specification;
7. the Drawings, and;
8. the Schedules and any other documents forming part of the Contract.

Words, Phrases and Expressions

routinely *adj*. 有规律地，通常地（regularly, not unusually)
integrate *vt*. 使…成为整体，融入
in general terms 概括地，笼统地（generally)
recital *n*. 一系列事件的详述
materialize *vt*. 具体化（to make something general into deails)
amplify *vt*. 详述（to explain in details)
administrative mechanisms 行政机制
starting date 开工日（commencement date)

date for possession 获得工地占有权的日期
completion date 竣工日
duration of the contract 工期
scheme design drawing 初步设计图，简图，草图
detail design drawings 详细设计图
as-built' drawings 竣工图
guideline *n.* 准则（rule）
ironmongery *n.* 铁器，金属件（ironware）
invaluable *adj.* 非常宝贵的（valuable）
dimensioned sketch 尺寸图
finish *n.* 表面光洁度，表面平整度，表面工艺水平
quantification *n.* 量化（being quantified）
systematic *adj.* 系统的
methodical *adj.* 有条理的
invariably *adv.* 始终不变的
particular *n.* 具体情况，详情
description *n.* 描述，形容
attendance *n.* 照管，看管，照顾
finalize *vt.* 最终定稿，最终敲定
contingency sums 偶发事件，意外事件费
distinctive *adj.* 有特色的，与众不同的
defective *adj.* 有缺陷的
practical completion 实际竣工，基本竣工（工程基本竣工后，工程师就可以向承包商颁发接收证书或移交证书）
final completion 最终竣工
non-completion 未竣工
schedule *n.* 资料表
setting-out *n.* 放线
discrepancy *n.* 差异（difference）
inconsistent *adj.* 不一致的（conflicting）
disruption *n.* 中断（halt）
govern *vt.* 管辖，控制（to dominate）
interpretation *n.* Notes（explanation）
noticeable *adj.* 明显的（obvious）
off the shelf 现成的（ready-made）
emphatically *adv.* 强调地
neutral *adj.* 中立的，不偏不倚的（impartial）
explanatory *adj.* Notes 的，说明的（interpretative）

categorically *adv.* 断言的，明确的（explicitly）

designate *vt.* 指明（to point out）

疑 难 词 句

1. Contract documentation is the means by which a designer's intentions are conveyed to the client, the statutory authorities, the quantity surveyor, the contractor and the sub-contractors.

译文：合同文件是将设计者的意图传达给顾客、执法部门、工料测量师、承包商以及分包商的一个手段。

2. However, all the items cannot come within the more limited legal definition of 'contract documents', a special term used in construction contracts to refer to those documents which define the contractor's obligations.

译文：但是，并不是所有的项目都能在这种具有更局限的法律定义的"合同文件"中找到，该"合同文件"是在施工合同中用来专指界定承包商义务的那些文件的专门术语。

3. However, there are different views between the forms of contract in whether or not such items as programmes, specifications and the like should be taken as contract documents.

译文：但是，不同形式的合同对于进度计划和技术规范等类的文件是否应被视为合同文件是有分歧的。

4. In defining the contractor's obligation, the articles do not descend to any great detail, but merely state, in a section starting with the word 'whereas', that 'the employer is desirous' of having the briefly described works carried out.

译文：在界定承包商的义务时，协议条款不会对此作详细描述，而只是在一段以"鉴于"开头的段落中陈述：业主希望这些被简单描述的工作得到实施。

Notes：Generally, the conditions of contract defines the contractor's obligations in great details, whereas, the articles of an agreement only give a very general idea of both parties' obligations.

5. They are also a useful control and check document to compare what has actually been done with what was originally planned.

译文：它们还是将实际完成的工作与计划完成的工作进行对比的一个有用的控制和检查文件。

6. The multiple uses which different drawings are required to perform, and the interaction between drawings and other types of documentation, have often been a source of problems.

译文：不同的图纸具有不同的用途，并且图纸与其他文件之间又有相互关联的作用，这又常常是问题产生的根源。

7. Assuming that such things are to be specifically controlled by the contract, rather

than relying on the terms which would be implied at common law, this is a task for the specification.

译文：如果这些事情可以被合同加以具体地控制而不是依赖于某些惯例或行规来约束的话，这就是技术规范的任务。

8. Not surprisingly, bills of quantity are increasingly being seen as hopelessly inadequate for all of these conflicting needs.

译文：并不令人惊讶的是，人们愈来愈认识到：工程量清单是没有希望满足所有这些相互矛盾的需求的。

9. To use the bill as the sole vehicle for both these purposes is an over-simplification that can lead to many problems.

译文：把工程量清单作为达到这两个目的的唯一手段有些过于简单了，会引起很多问题。

10. These may be necessary for the contractor to execute the work but such additional information must not impose extra work on the contractor.

译文：这些文件对于承包商履行合同任务可能是必要的，但这些增加的资料不能作为额外的工作强加给承包商。

11. Although the contractor must provide a programme immediately after being awarded the contract, no extra obligations should be imposed on the part of contractor to do it before a certain legally binding date.

译文：虽然承包商必须在授标后马上提供进度计划，但在某个法定日期前不应给承包商增加任何义务。

12. That rule is based on the sensible idea that documents which are prepared for a specific job, rather than being taken 'off the shelf', are more likely to reflect the parties' true intentions.

译文：这个规则是基于一个合情理的观点，即：专门为一个具体的工作准备的文件比现成的文件更可能反映各方真实的意图。

13. If it were to be applied to a building contract, it would mean for example that provisions in bills of quantity would override the printed conditions of contract.

译文：如果这个规则用于建设合同，比方说工程量清单的条款就会优先于印制的合同条件。

5.2 Implied Terms

Construction contracts are, like other contracts for work and materials, subject to the implied terms contained in the law of contract. In addition, where the contract document itself is not detailed, explicit, or exhaustive, it may be possible for terms to be implied by the courts into such contracts, if necessary.

In considering this question, it is important to realize that two different kinds of implication may be involved. First, certain terms will be automatically implied as a matter of law into certain classes of contract, provided only that they are not inconsistent with any express terms. Secondly, and far less common, a term may be implied as a matter of fact into an individual contract where that contract would be commercially unworkable without it. Again, of course, there can be no implication of a term which would conflict with an express term.

These two kinds of implication may be separately considered as below.

5. 2. 1 Implication in law

The law regards certain terms as 'usual' in building contracts, so much so that, if the contract is silent on these matters (as may well be the case with small informal agreements), these terms will be implied. However, it must be stressed once again that the court will not rewrite a contract freely entered into, and so there will be no implication of terms which would be inconsistent with the express agreement. This will have a direct practical effect: where, as under the main standard-form building contracts, all the 'usual' matters are covered in some detail, implied terms are not applicable.

The terms which are regarded as 'usual' in this respect, and which will therefore be implied into construction contracts, will not be dealt with in details here. Nevertheless, a brief summary of the most important terms may be useful at this stage:

- *Employer's obligations*: The implied obligations of the employer, though as many as can appear on a list, can effectively be reduced to two: a general duty not to hinder the contractor's efforts to complete the work and, more positively, a duty actively to co-operate with the contractor. The former branch has been held to be breached, for example, by an employer causing delay via his employees or agents (who include, for this purpose, the supervising Engineer, the contract administrator, but not independent third parties or even nominated subcontractors); interfering with the supply of necessary materials to the contractor; and interfering with the Engineer's function as independent supervisor under the contract.

As to the duty of positive co-operation, this involves such matters as giving possession of the site; appointing an architect and nominating sub-contractors and suppliers; supplying the necessary instructions, information, plans and drawings; and, if the Engineer or contract administrator persists in managing the contract in a unfair manner, dismissing him or her and appointing a replacement. All these things must be done and, what is more, they must all be done without undue delay.

- *Contractor's obligations*: Where a building contract does not specify a date for completion, or where the employer has delayed completion, the contractor's implied duty is to complete within such a time as is reasonable in all the circumstances.

As to the standard of the work, it is implied that the contractor will use proper workmanship and, in most cases, that the materials used will be of good quality and fit for their purpose. However, the 'fitness for the purpose' warranty will not be on the part of contractor where it is clear that the employer has not relied on the contractor 's 'skill and judgment' in selecting the materials, such as where the employer specifies the material and nominates the supplier. Further, there may be circumstances in which even the warranty as to quality will not be implied, for example where the employer knows that the contractor will have no right of claim against the actual supplier of the materials.

Apart from the fitness for the purpose of the materials used, there may in certain cases be an implied warranty that the completed works will be fit for their purpose. Such a term depends upon the employer's reliance on the expertise of the contractor. Accordingly it cannot apply where the contractor is merely to build in accordance with detailed plans and specifications. Similarly, it cannot apply where the contractor is under the supervision of an architect. A warranty of fitness will be fairly readily implied, for example, into a contract to buy a house which is in the course of being built. However, the sale of an already completed building is not subject to such implication, in this case the buyer should be self-responsible for examining its quality.

5.2.2 Implication in fact

As we knew earlier that a court will try to make sense of the parties' agreement, for example in identifying the form of contract to which they have mistakenly referred. However, the courts have repeatedly stated that they will not make or improve contracts for the parties. The principle of freedom of contract means that, if parties have entered into a contract which is unreasonable, inconvenient or commercially unwise, it is not for the courts to change their arrangement. They must bear the consequences themselves. A term will not be implied into a particular contract just to correct or amend the contract to make it more reasonable, more convenient, or more commercially wise; it will only be implied if the missing implication is so glaringly obvious that both parties must have intended to include it.

Words, Phrases and Expressions

implied term 默认条款
implication *n.* 含义，暗指
interfere with 干预，阻扰
positive *adj.* 积极的，肯定的
replacement *n.* 替换的人
undue *adj.* 不当的，过分的
reliance *n.* 依靠，依赖

疑 难 词 句

1. In addition, where the contract document itself is not detailed, explicit, or exhaustive, it may be possible for terms to be implied by the courts into such contracts, if necessary.

译文：此外，当合同文件本身不详细，不清楚，或不全面时，如果有必要可以将法院的默认条款作为这些合同的条款。

2. Secondly, and far less common, a term may be implied as a matter of fact into an individual contract where that contract would be commercially unworkable without it. Again, of course, there can be no implication of a term which would conflict with an express term.

译文：第二种情况比第一种情况少见：当某个合同因缺少某个事实而变得在商业上无法操作时，这个事实就可以作为该合同的默认条款。当然，这个默认条款同样也不能与明示条款相冲突。

3. This will have a direct practical effect: where, as under the main standard-form building contracts, all the 'usual' matters are covered in some detail, implied terms are not applicable.

译文：这就会引起一个直接的实际效果：如果一些主要的标准格式的建设合同已经比较详细地涉及到了所有的"惯常"事件，默认条款将不适用。

4. However, the 'fitness for the purpose' warranty will not be on the part of contractor where it is clear that the employer has not relied on the contractor ' s 'skill and judgment' in selecting the materials, such as where the employer specifies the material and nominates the supplier.

译文：但是，当业主选择材料时并未依赖承包商的"技术和判断"的事实很清楚时，比如业主规定了材料并指定了供货商，这种情况下就不应该要求承包商作出"适合用途"的保证。

5. It will only be implied if the missing implication is so glaringly obvious that both parties must have intended to include it.

译文：只有当双方包含这项缺失的默认条件的意图非常明显时才能将它作为默认条件。

Chapter 6 Contractor's Obligations

First of the contractor's obligations is the obligation to construct the works in accordance with the contract documents within the required time. Here, we will discuss the fundamental obligations and associated obligations as to workmanship, quality of materials, co-ordination and management of the site. Obligations can arise both in tort and in contract concurrently.

The exact work scope of what the contractor undertakes depends upon the wording of the contract. If the contract is for a complete building, such as a residential house, then it is implied that the contractor will provide everything which is indispensably necessary to achieve the result, regardless of what is specified in the specifications. Where the contract is clear that the bills of quantity contain an exhaustive definition of the works, the contractor's obligation is limited to providing what is described in the bills. If this is not sufficient for completion of the building, the employer must pay for extra work.

6.1 Standard of work

The contractor's basic obligation, so far as the standard of work is concerned, is to comply with the terms of the contract. This includes both express terms (such as the requirement that work shall be of the standards described in the bills or specifications) and implied terms (such as the principle that all materials shall be of 'satisfactory quality').

In addition, there are often provisions in some contracts for work to be to the satisfaction of the Engineer. This might seem to suggest that the Engineer has an absolute right of rejection of such work, but meanwhile the contracts may place a limit upon this power, for example stating that, where approval of workmanship or materials is a matter for the opinion of the contract administrator, 'such quality and standards shall be to the reasonable satisfaction of the contract administrator'. Therefore, when the work is subject to the subjective view of contract administrator's satisfaction, the wording of a contract apply reasonableness to the satisfaction. However, some other contracts might be contrasted with it in which the Engineer may have complete authority to approve or reject work, regardless of what the bills might say about quality or workmanship, with no apparent test of reasonableness.

Where a contract requires work to be to the satisfaction of the contract administrator, an important issue is whether this becomes the only contractual standard, or whether the contractor must also satisfy whatever objective standards are contained, expressly or impliedly, in the contract documents. In other words, is it open to an employer to argue that, notwithstanding its approval by the contract administrator, certain work does not comply with a standard specified in the contract requirements, or falls below the standard of 'satisfactory quality' laid down in a certain governmental regulation? In short, is the contractor under one obligation or two?

The answer to this question, as is so often the case, is that everything depends upon the wording of the particular contract, although the weight of authority has traditionally been in favour of making the contractor satisfy both standards.

The phrase 'standard of work' in fact stands for three separate, though linked, ideas, which are discussed below.

6.1.1 Workmanship

The standard of workmanship may be defined in considerable detail by the contract, for example by requiring it to comply with an appropriate code of practice. Under some contracts, such a requirement would appear in the bills or in the specifications. For example, it may state that 'all workmanship shall be of the standards described in the contract bills', then the clause may go on to state that, where the bills do not set a standard, the workmanship 'shall be of a standard appropriate to the works'. This would in any case be implied by law since, in the absence of any express terms covering this issue, the courts will imply a term in the contract that the work will be carried out with proper skill and care, which means in a 'workmanlike' manner.

It is at least arguable that 'workmanship' refers only to the standard of the finished item, and not to the method used to achieve that standard, certainly in building contracts. If this is so, then the contract administrator has no power to control the manner of working at the time it is being carried out. However, many contracts give the contract administrator such power by providing that 'all work shall be carried out in a proper and workmanlike manner'. If this is breached by the contractor, then the contract administrator may issue whatever instructions are necessary. Even if these constitute a variation, the contractor will not be entitled to any payment nor to any extension of time. By contrast, other contracts may not have such a strict limit to the methods of work.

It is important to note that the contractor is responsible, not only for personally performing unsatisfactory workmanship, but also for that of any sub-contractor or agents.

6.1.2 Standard of materials

Some contracts state that all materials and goods shall, so far as procurable, be of the respective kinds and standards described in the contract bills or specifications. This wording allows for the eventuality that some things which were available at the time of tender may become unavailable. Some other contracts, however, place a similar obligation upon the contractor, but none of them makes allowance for materials which may have become unprocurable.

These express obligations are backed up by an implied term requiring all goods and materials supplied to be of 'satisfactory quality'. This phrase, which replaced the time-honoured expression 'merchantable quality', means that they are to be as free from any defects as it is reasonable to expect, considering such factors as their price and the way they are described.

The law of contract in many countries state that a contractor will be liable if materials are unsatisfactory, even where those materials have been selected by the employer (e.g. by nominating a particular supplier). However, the contractor will not be liable for defective materials where forced by the employer to procure those materials from a supplier who, to the employer's knowledge, excludes or limits liability for defects. This last principle is now expressly provided for by many international contracts.

6.1.3 Suitability of materials

Even where materials are 'of satisfactory quality' in the sense of being free from defects, they may still not be fit for the purpose for which they are used where this is the case, it is again possible for the contractor to be liable, but the term which is chosen to be implied here is of a more limited kind, that means it cannot be used in any case. For example, a certain code of practice implies a term that goods shall be reasonably fit for the purpose for which they are supplied, but only where it is clear that the recipient is relying on the skill and judgment of the supplier. In practical terms, it is only when the contractor has the choice of materials that liability can be imposed on the contractor if what is selected is unfit for its purpose. Therefore, there will be no such implied term in respect of materials which are specified by the employer or the architect, for in this case there is no reliance on the contractor.

6.1.4 Suitability of the building

Besides the obligation to provide suitable materials, a contractor may in some circumstances be subject to an implied term that the building itself, when completed, will be fit for its intended purpose. Such a term, which will normally be implied only into 'design and build' contracts.

Words, Phrases and Expressions

fundamental obligation 基本义务

associated obligation 关联义务

tort *n.* 民事侵权行为

concurrently *adv.* 同时地（simultaneously）

undertake *vt.* 承担，承揽（to assume）

exhaustive *adj.* 全面的，彻底的（all-around，all-sided）

provision *n.* 条款，规定（article）

rejection *n.* 拒绝（refusal）

code of practice 实施代码，实施规程

workmanlike *adj.* 精工细作的（professional）

arguable *adj.* 可辩论的（controvertible）

finished item 已完成的项目，成品

procurable *adj.* 可得到的（available）

eventuality *n.* 不测事件（unexpected event）

satisfactory quality 令人满意的质量

time-honoured *adj.* 由来已久的

suitability *n.* 合适，适合（fit）

疑 难 词 句

1. However, some other contracts might be contrasted with it in which the Engineer may have complete authority to approve or reject work, regardless of what the bills might say about quality or workmanship, with no apparent test of reasonableness.

译文：但是，另外一些合同却有与此相反的规定，即：工程师可以不必考虑工程量清单对质量或工艺水平的规定而对工程行使全权批准或拒绝的权力，对其合理性没有明显的要求。

2. In other words, is it open to an employer to argue that, notwithstanding its approval by the contract administrator, certain work does not comply with a standard specified in the contract requirements, or falls below the standard of 'satisfactory quality' laid down in a certain governmental regulation?

译文：换句话说，业主是否可以提出这样的质疑，即：尽管工程获得了合同管理者的批准，但某些工程并不符合合同规定的标准，或低于某些政府规定中的“令人满意的质量”标准？

3. This wording allows for the eventuality that some things which were available at the time of tender may become unavailable.

译文：该措辞接受这种不测事件，即：某些物资投标时是可以获得的，而随后却可能变得不可获得了。

4. This phrase, which replaced the time-honoured expression 'merchantable quality', means that they are to be as free from any defects as it is reasonable to expect, considering such factors as their price and the way they are described.

译文：该措辞取代了由来已久的“适合买卖的质量”的说法，其含义是：所提供的物资和材料不应有根据产品价格和对产品的描述而能够合理预期不应有的任何瑕疵。

5. The law of contract in many countries state that a contractor will be liable if materials are unsatisfactory, even where those materials have been selected by the employer (e. g. by nominating a particular supplier).

译文：许多国家的合同法都规定：如果材料不满意，承包商应该负责，尽管这些材料是业主选定的（比如通过指定具体的供货商）。

6. For example, a certain code of practice implies a term that goods shall be reasonably fit for the purpose for which they are supplied, but only where it is clear that the recipient is relying on the skill and judgment of the supplier.

译文：比如，某个实施代码的默认规定是所供货物必须合理地适合提供该货物的目的，但条件是货物的接受者是依赖供货商的技术和判断购买的这些货物。

6. 2 Statutory obligations

Statutory obligations are the obligations imposed by some legislations such as government regulations. All parties must of course obey the law and are thus bound by these obligations. However, failure by the contractor to comply with a statutory obligation would not in itself amount to a breach of contract for which the employer could claim damages. That will only be the case where the contract specifically requires the contractor to comply with the relevant statutory rules.

The reason for making the statutory compliance a contractual requirement is that, where the contract works are carried on in such a way as to breach a statute, the effect may well be that the employer becomes liable to criminal prosecution or to civil actions brought by third parties who are adversely affected. By placing a contractual obligation of compliance on the contractor, the contract seeks to ensure that, in such circumstances, the employer will be entitled to claim an indemnity from the contractor.

6. 2. 1 Contractor's duties

Many local and central governments in many countries have promulgated decrees, codes, bylaws, or regulations controlling building work for contractors to obey. But the contractor's obligation extends much wider than these. Further, contractors should also obey all the regulations to which the works are to be connected.

In relation to any fees, charges, rates or taxes which become payable in respect of the works, usually the relevant laws place the primary obligation of paying these upon the contractor. However, the ultimate cost will be borne by the employer, for the amounts are to be added to the contract sum.

6.2.2 Divergence between statutory requirements and contract

Usually, if the reason why there is a breach of statute is that the work, as designed, does not conform to statutory requirements, then responsibility for this falls upon the employer and not upon the contractor. The contractor's only obligation is to notify the contract administrator of any apparent divergence between the contract documents and the statutory requirements. Provided the contractor does this, the contract administrator must then within seven days issue instructions so as to bring the works into line with the statutory requirements. Such instructions will be treated as a variation, which means that the contractor will be paid for extra work. Further, where compliance with the instructions causes delay or disruption, the contractor will be entitled to claim an extension of time and for loss and/or expense.

The treatment of divergences described here applies to most contracts. However, some form contracts are very different, stating that except where the employer's requirements are specifically stated to be in accordance with statutory requirements, it is the contractor's responsibility to ensure that the work as designed and built does not disobey the statute. As a result, any divergence must be put right at the contractor's own expense.

6.2.3 Emergency work

Sometimes, emergency work is necessary in order to comply with statutory requirements. This is likely to occur where work is necessary to ensure the continuing safety of people and property; for example, where a neighbouring property has become unsafe and may collapse on to the site. The contractor must immediately (i.e. without waiting for instructions from the contract administrator) undertake such limited work and supply such limited materials as are necessary to ensure compliance with statutory obligations. Provided that the emergency is then shown to have arisen because of a divergence between the contract documents and the statutory requirements, the contractor's work will be valued as a variation.

However, some other contracts do not mention emergency work at all, which suggest that the cost of such work is to be borne entirely by the contractor. Where the emergency arises because of a divergence between the contract documents and statutory requirements, the engineer must issue a variation instruction and the contractor will be paid for this work. However, even in this case it is arguable that the contractor will remain uncompensated both for the earlier abortive work and for the cost of removing it.

Words, Phrases and Expressions

statutory *adj.* 法定的（legal）
legislation *n.* 法律，法规（law，statute）
breach *vt.* 违反（to disobey）
claim *vt.* 索赔
statute *n.* 法令，法规
criminal prosecution 刑事检控
civil action 民事诉讼
adversely *adv.* 不良地，负面地
indemnity *n.* 赔偿，补偿（compensation）
promulgate *vi.* 颁布（to issue）
divergence *n.* 分歧（difference）
emergency work 应急工程
collapse *vi.* 倒塌
abortive *adj.* 失败的

疑 难 词 句

1. However，failure by the contractor to comply with a statutory obligation would not in itself amount to a breach of contract for which the employer could claim damages.

译文：但是，承包商不遵守法律义务的行为本身并不一定会构成业主可以提出索赔的违约。

2. Further，contractors should also obey all the regulations to which the works are to be connected.

译文：此外，承包商还应该遵守与本工程相关的所有规章和规则。

3. However，some form contracts are very different，stating that except where the employer's requirements are specifically stated to be in accordance with statutory requirements，it is the contractor's responsibility to ensure that the work as designed and built does not disobey the statute.

译文：但是，有些格式合同对此却有不同的规定，即：除非合同明确说明业主要求是符合法律要求的，否则，确保工程的设计和建造不会违背法律是承包商的责任。

Notes：It means that the contractor is liable for putting right the divergence between the employer's requirements and the statutory requirements at his own expense.

4. Provided that the emergency is then shown to have arisen because of a divergence between the contract documents and the statutory requirements，the contractor's work will be valued as a variation.

译文：只要该应急工程随后被证实是由合同文件与法律要求相冲突造成的，承包商的该应

急措施将被作为变更予以估价。

5. However, even in this case it is arguable that the contractor will remain uncompensated both for the earlier abortive work and for the cost of removing it.

译文：但是，即使是在这种情况下，承包商以前干的错误工程和拆除该工程的费用仍然得不到赔偿，这是个有争议的问题。

6.3 Co-ordination and Management

Apart from the main contractor's role as builder (which may be very limited where extensive use is made of sub-contractors, or even non-existent under a management contract), the main contractor has an important part to play in managing the site and the persons who work on it. This covers a number of issues, which are dealt with below.

6.3.1 Control of persons on the site

It is the main contractor's responsibility to programme the overall project and to co-ordinate the contributions made by various other persons and organizations, notably sub-contractors.

Apart from sub-contractors, an important group of participants in construction projects consists of statutory undertakers. There are two types of these: those who exercise some sort of power over the works, and those who have to undertake work on the site. Examples of the first type are the public bodies which exercise supervision over the design, quality or safety of the works by the enforcement of planning legislation or building regulations. Examples of the second type are those public bodies or private firms which provide utility services such as gas, water, electricity and drainage. With the latter, the contractor must co-ordinate and connect with their work, but will not exercise the same powers of control over their work as over that of sub-contractors.

In some countries it is often the case that statutory undertakings will undertake work which is not part of their statutory powers. Similarly, private firms may undertake work which is subject to statutory regulation and also work which is not. For example, an electricity board may be employed to carry out sub-contract works. In this case, they are not treated as a statutory undertaking, but are in the same position as any other sub-contractor (nominated or domestic). It is obviously of great importance to know in what capacity a statutory undertaking is operating in any particular case, but unfortunately this is not always clear, especially where they exceed their powers.

Although overall responsibility for the way in which work is carried out lies with the contractor, the contract administrator has a supervisory role to play. To enable this to take place, many contracts stipulate that the contractor must allow the contract administrator to have access to the works and to the contractor's workshops as well as sub-contractors' workshops at all reasonable times. Furthermore, if the employer appoints a person to act as inspector under the directions of the contract administrator, the contractor should provide every reasonable facility for the performance of that duty.

It may finally be noted, as a practical feature of the contractor's responsibility for management and co-ordination, that many contracts usually specifically require the contractor to always keep 'a competent person-in-charge' on the work site.

6.3.2 Exclusion of persons from the works

All the main standard-form contracts give power to the contract administrator to order the exclusion from the works of specified persons. And this authority sometimes or under some contracts can include not only the contractor's own employees also include the employees of sub-contractors. However this authority cannot be exercised 'unreasonably or vexatiously'. Its purpose is to exclude persistent bad performers, although this is not made explicit. However, the misuse of this provision would occur, in this case the question of unreasonableness, or vexation, is one which can only be decided by an arbitrator.

Most standard-form contracts require the contractor to employ only those people who are properly skilled and undertake their duties with an appropriate level of care. The contract administrator or engineer have the power to order the removal from site of any of the contractor's employees, but only for lack of diligence or workmanship.

According to the provision in this respect of some contracts, the project manager of the employer may have extremely far-reaching powers to exclude any individual from the site. No reason need be given, and the contractor cannot question the exclusion. This use of the power has usually nothing to do with workmanship and diligence, and appears to be a matter of security.

Under some construction-concerning regulations in many countries, the contractor must exclude all unauthorized persons from the site. This additional requirement applies to all construction work governed by the regulations and it places a very strict obligation on a contractor to maintain proper security and careful governance of admissions to site. This particular regulation carries civil as well as criminal liability.

6.3.3 Antiquities

All the standard-form contracts have the provisions dealing with the discovery of interesting objects, new or old, during the course of the work. One important reason is that the removal or disturbance of such objects may actually destroy their worth. One example of this was the discovery of Shakespeare's Globe Theatre in London, where an archaeological find of great importance delayed a very large and expensive development project for a considerable period of time. More generally, fossils may be valuable, and some types may completely degrade on exposure to atmosphere for a length of time. Further, one of the most significant features of a fossil is the exact location in which it is found. It must be left in place, because a random fossil is of little use to a paleontologist without data about its location. A third problem relates to the ownership of treasure trove.

Most standard-form construction contracts contain the provisions to cover these issues, the wording is almost equivalent to following: "*All fossils, antiquities and other objects of interest or value which may be found on the site or in excavating the same during progress of the Works shall become the property of the employer.*" For example, FIDIC Conditions of Contract for Construction states "*All fossils, coins, articles of value or antiquity, and structures and other remains or items of geological or archaeological interest found on the site shall be placed under the care and authority of the Employer. The Contractor shall take reasonable precautions to prevent Contractor's Personnel or other persons from removing or damaging any of these findings.*"

6.3.4 Testing and approvals

The question of whether or not a contractor's work is up to the contractual standard may arise in arbitration or litigation, in the course of a claim for damages for breach of contract. More frequently, however, it will be something for the contract administrator to decide through the certification procedure. Whatever the precise nature of the contract administrator's role in approving work, many contracts oblige the contractor to open up or uncover work for inspection or testing. Such clauses usually make provision for the contractor to be paid compensation in the event that no defects are found.

Words, Phrases and Expressions

participant *n.* 参与者
statutory undertaker 执法人员
enforcement *n.* 实施，执行（execution）
planning legislation 规划法规
building regulation 房建规则
utility *n.* 公用事业
inspector *n.* 检查员，检验员
exclusion *n.* 排除
vexatiously *adv.* 厌烦地
persistent *adj.* 持续的，不断的

arbitrator *n.* 仲裁人
diligence *n.* 勤勉，勤奋
far-reaching *adj.* 深远的
unauthorized *adj.* 未被授权的
governance *n.* 统治，治理
archaeological *adj.* 考古的
degrade *vi.* 使降级
random *adj.* 任意的，随机的
paleontologist *n.* 古生物学者
treasure trove *n.* 无主埋藏物
litigation *n.* 诉讼
oblige *vi.* 迫使做

疑 难 词 句

1. Examples of the first type are the public bodies which exercise supervision over the design, quality or safety of the works by the enforcement of planning legislation or building regulations.

译文：第一种类型的例子是公共机构，它们通过实施规划法规或房建规则对工程的设计、质量或安全行使监督权。

2. It is obviously of great importance to know in what capacity a statutory undertaking is operating in any particular case, but unfortunately this is not always clear, especially where they exceed their powers.

译文：知道某个法规的实施在任何一个具体情况下的操作能力显然是非常重要的，但遗憾的是这并不总是很清晰，尤其是当它们超越其权限的时候。

3. This additional requirement applies to all construction work governed by the regulations and it places a very strict obligation on a contractor to maintain proper security and careful governance of admissions to site.

译文：这一额外要求适用于该规定管辖下的所有的建设工程，并且还赋予了承包商一个非常严格的义务，即：承包商必须保持工地的适当安全并对进入工地实行严格的管理。

4. Such clauses usually make provision for the contractor to be paid compensation in the event that no defects are found.

译文：这些条款通常规定：如未发现瑕疵，承包商应该得到赔偿。

6.4 Transfer of materials

The common law rules governing the transfer of ownership in materials from contractor to employer are fairly straightforward. They will operate as implied terms in any contract unless they are otherwise specified by express provisions. Such express provisions, known as retention of title clauses, are sometimes found in sub-contracts or suppliers' standard terms.

6.4.1 General position

According to common law, as soon as any materials or goods are incorporated into a building, they cease to belong to the contractor and become the property of the employer. Until the materials are built in to the works, however, even though they have been delivered to site, they remain the property of the contractor. Even the fact that the employer has paid the contractor for the materials will not transfer their ownership to employer, unless (as is usually the case) the contract makes express provision for this.

The question of whether ownership lies with the employer or the contractor becomes most important where one of the parties becomes insolvent. As a general principle, all the creditors of an insolvent person or company are entitled to share equally in the remaining assets, and these will include all property which is owned by the insolvent person at the time. Thus, if the contractor becomes insolvent, materials which are on site but which have not yet passed to the employer may become the property of the contractor's receivers. What is more, unless the contract provides otherwise, the ownership of the materials will still be transferred to the contractor's creditor by law even where the value of those materials has been included in interim certificates.

Somewhat strangely, the courts have not always applied the same principle to cases where it is the employer who becomes insolvent. It has for instance been held that, although the ownership of building materials may have passed to the employer under a 'vesting clause', the contractor's continuing rights to use those materials in constructing the building are strong enough to override the claims of the employer's creditors.

6.4.2 Contract provisions

Some contracts do not alter the principle that the ownership of materials is transferred when they are incorporated into the building, but some other contracts provide a method by which the ownership may pass to the employer at an earlier stage. For example, under some contracts, the value of unfixed materials and goods intended for the works, whether they are on or off site, may be included in an interim certificate. If this is done, then the ownership of those materials and goods will pass to the employer as soon as the amount is duly paid.

Besides, some other contracts depart further from the common law position. For example, a contract states baldly that 'All Contractor's Equipment Temporary Works materials for Temporary Works or other goods or materials owned by the Contractor shall when on Site be deemed to be the property of the Employer'. Instead of this rather drastic statement there are still some milder provisions that the contractor shall have exclusive use of all the materials; that the contractor may remove the materials when they are no longer required for the works; and that the ownership of the materials so removed shall then revert to the contractor. As for materials which have not been delivered to the site but which are ready for incorporation in the works, a certain provision enables the contractor to transfer the ownership to the employer, and then, after engineer gives his written approval to the materials, the contractor is entitled to be paid for them.

6.4.3 Retention of title

The contract may deal satisfactorily with questions of ownership as between the employer and the contractor; however, problems may arise in cases where the contractor brings on to the site materials which are still in the ownership of a supplier. In particular, many suppliers operate under standard conditions of sale, which stipulate that they shall retain the ownership of goods until full payment is made. Therefore, there is here an obvious possibility of conflict between the terms of the main contract and those of the contract of supply.

Words, Phrases and Expressions

transfer *vt.* 转移 (to shift)
straightforward *adj.* 简单的，易懂的 (simple)
title *n.* 权益，所有权 (ownership)
common law 习惯法，惯例 (practice)
incorporate *vt.* 使并入
creditor *n.* 债权人
receiver *n.* 接收人，接管人，财产管理人
vesting clause 归属条款
override *vi.* 优先于，压倒 (to govern)
unfixed material *adj.* 未固定的材料（在此指"还未被工程使用的"建造材料）
depart from 违背，背离 (to disobey)
baldly *adv.* 坦率地 (frankly)
drastic *adj.* 严厉的，激烈的 (furious)
milder *adj.* 温和的 (gentle)
exclusive use 专用
revert to 归还
deal with *vt.* 处理

疑 难 词 句

1. They will operate as implied terms in any contract unless they are otherwise specified by express provisions.

译文：在任何合同中它们都可以作为默认条款来操作，除非合同中有明示条款对此另有规定。

2. Even the fact that the employer has paid the contractor for the materials will not transfer their ownership to employer, unless (as is usually the case) the contract makes express provision for this.

译文：即使业主已经向承包商支付了材料款，材料的所有权也不能因此转移给业主，除非（通常情况是这样）合同对此有明示条款。

3. What is more, unless the contract provides otherwise, the ownership of the materials will still be transferred to the contractor's creditor by law even where the value of those materials has been included in interim certificates.

译文：更有甚者，即使材料的价值已经包括在期中付款证书里，材料的所有权仍然会根据法律移交给承包商的债权人，除非合同另有规定。

4. It has for instance been held that, although the ownership of building materials may have passed to the employer under a 'vesting clause', the contractor's continuing rights to use those materials in constructing the building are strong enough to override the claims of the employer's creditors.

译文：例如，曾经有这样的判例：虽然建筑材料的所有权根据"归属条款"可以移交给业主，但承包商在施工过程中对这些材料的继续使用权的地位足以优先于业主债权人的索赔。

5. For example, under some contracts, the value of unfixed materials and goods intended for the works, whether they are on or off site, may be included in an interim certificate.

译文：例如，有些合同规定：拟用于工程但还未被工程所使用的材料和物资，无论它们是否在工地，其价值都可以包括在期中付款证书中。

6. Therefore, there is here an obvious possibility of conflict between the terms of the main contract and those of the contract of supply.

译文：因此，总承包合同和供货合同之间发生冲突的可能性是显而易见的。

Chapter 7 Employer's Obligations

As we all know, the most important of the employer's obligations under a construction contract are monetary: to pay the contractor what is due for work done, and in certain circumstances to compensate the contractor for loss and expense. These two major obligations are basic and easy to understand.

Here, we are going to discuss a number of other obligations which are imposed on the employer. We look at those obligations which will be implied by law wherever they are not overridden by some express term of the contract.

The reason why the employer appears to have relatively few express contractual obligations, and to play a merely passive role, is that the contract allocates numerous duties to the contract administrator (or Engineer under FIDIC conditions). In truth, many of these duties are the employer's responsibility in the sense that, if the contract administrator fails to perform, the contractor may claim against the employer for breach of contract.

As many of the employer's obligations are carried out by the contract administrator or Engineer and meanwhile in fact some employers always try to communicate directly with the contractor, which would lead to a serious risk of causing confusion and resulting problems. Therefore a wise contractor will decline to accept direct instructions from the employer, requesting instead that they be sent through the contract administrator. To avoid problems, in any event, the contractor should always ask the contract administrator for written confirmation of the employer's direct instructions.

7.1 Implied obligations

It is obvious to everyone that the contractor has a duty to carry out and complete the contract works. What is sometimes overlooked, however, is that the contractor also has a right to do this. Unless the contract provides otherwise, the contractor is entitled to carry out the whole of the contract works within the contract period, and the employer must co-operate to enable this to be achieved.

7.1.1 Non-hindrance and co-operation

Usually, two general obligations on the part of the employer are to be implied into all building contracts. They are expressed as follows:

1. The employer will not hinder or prevent the contractor from carrying out all obligations in accordance with the terms of the contract, and from executing the works in a regular and orderly manner.

2. The employer will take all steps reasonably necessary to enable the contractor to fulfill all obligations and to execute the works in a regular and orderly manner.

These two employer's obligations, though expressed separately, are really the positive and negative aspects of the same thing. Together they are the same employer's duty to co-operate with the contractor in all aspects of the contract work. In considering what this means in practice, accordingly we do not have to divide our examples into those which involve 'non-hindrance' and those which require 'co-operation'.

7.1.2 Specific examples of non-hindrance and co-operation

One aspect of the contractor's right to do the work tendered for is that the employer cannot order the omission of work from the contract, with the intention of giving this work to another contractor. This indeed is so fundamental that such conduct could well constitute a serious breach by the employer, which would entitle the contractor to terminate the contract. And similarly the employer cannot take work for which the main contractor has priced and then nominate a sub-contractor to carry it out.

The contractor's right to carry out the work also lies in that some contracts such as JCT 80 provide that the employer cannot unilaterally abandon the project altogether. However, some other standard form contracts (such as MC 87 and FIDIC Conditions of Contract) do contain an express clause allowing the employer to terminate the project at any time without reason. Naturally, under such a contract, an employer who abandons the project will not be liable for all the damages to the contractor. Therefore it is important for a contractor to ensure that, where the main contract contains a unilateral termination clause, the sub-contract contains a similar provision. If this is not done, the contractor will be liable to the sub-contractor when the project is abandoned and will not be able to pass responsibility for this on to the employer.

Similarly, as with abandonment, the same question is also with temporary suspension of the work, although it is far more common to find an express term of the contract giving the employer certain powers in this respect.

Further, the employer's duty of co-operation involves giving the contractor possession of the site at the time specified in the contract. If the contract is silent on this matter, then the employer's implied obligation requires the contractor to be given possession at such a time as will enable the work to be finished by the specified completion date.

Whether or not the employer's obligation extends to the obtaining of any official permits for the proposed work, such as planning permission, depends essentially on the terms of the contract. Under JCT 80, for example, that the employer should be responsible for ensuring that the designed works comply with all statutory requirements; thereafter during the fabrication phase, however, the contractor should be responsible for obtaining whatever permissions. By contrast, many design and build contracts place the entire responsibility on the contractor.

An important part of the employer's duty of co-operation concerns the appointment of a contract administrator or engineer under some contracts and, where appropriate, the nomination of sub-contractors. These matters are usually covered by express terms of the contract but, if not, the implied obligation certainly concerns them. Indeed, where the contract envisages the use of an architect or an engineer to supervise the works and act as the employer's agent, the appointment thereof should be a condition precedent to the performance of the contractor's obligations. This means that a contractor can refuse to carry out any work until the employer makes such an appointment. If, however, the contractor proceeds without the appointment of a contract administrator, then this would release the employer from the responsibility of such an appointment.

Similarly, where the contract requires the sub-contractors or suppliers to be nominated by the employer, the employer must do it within a reasonable time. Further, should it become necessary to renominate (or to reappoint a contract administrator), this too must be done within a reasonable time.

As far as the general operation of the project is concerned, the employer may be put in breach of the duty of co-operation by what is done by a number of other parties. For example, the employer will be liable to the contractor for breach of contract if the contract administrator delays unreasonably in giving necessary instructions; if materials which the employer has undertaken to supply are delivered late; or if the contractor's work is impeded by other contractors working directly for the employer. However, any responsibility for the defaults of nominated sub-contractors or suppliers must be based on an express term of the contract, namely it is not necessarily imposed on the part of employer.

Still as for the subject of the employer's implied obligations, it is important to note that these do not require co-operation to an extent which enables the contractor to do more than the contract specifies. Thus, while it is the contractor's right to complete the project on or before the contractual completion date, the employer's duty of co-operation extends only to ensuring that the completion date is achieved. A contractor who seeks to finish early cannot force the employer to assist. Similarly, where a sub-contractor undertakes to carry out work 'at such time or times as the contractor shall direct or require', there is no implied term requiring the main contractor to make sufficient work available to the sub-contractors to enable them to work in an efficient and economic manner.

Words, Phrases and Expressions

monetary *adj.* 金融的，货币的（financial）
compensate *vt.* 补偿，赔偿（to pay for）
overlook *vt.* 忽视（to neglect）
hinder *vt.* 阻碍，妨碍（to impede）
orderly *adj.* 有序的（well-arranged）
omission *n.* 省略，删减（being left out）
terminate *vt.* 终止（to stop，to end）
provide *vt.* 规定（to stipulate）
unilaterally *adv.* 单方面的，单边的
abandonment *n.* 放弃（giving up）
envisage *vt.* 设想（to assume）
precedent *adj.* 优先的，先决的
proceed *vi.* 进行
release *vt.* 释放，解除（to free from）
impede *vt.* 阻碍，妨碍（to hinder）

疑 难 词 句

1. We look at those obligations which will be implied by law wherever they are not overridden by some express term of the contract.

译文：我们将了解那些法律默认并且同时又不会被合同的某些明示条款改写（不附属于明示条款）的义务。

2. In truth，many of these duties are the employer's responsibility in the sense that, if the contract administrator fails to perform，the contractor may claim against the employer for breach of contract.

译文：事实上，这些职责中的许多职责都是业主的责任，因为如果合同管理者不能履行这些职责，承包商可以向业主提出违约索赔。

Notes：in the sense 相当于 in the way，中文意思是：在……意义上；以……方法；由于……

3. These two employer's obligations, though expressed separately，are really the positive and negative aspects of the same thing.

译文：这两个业主义务尽管是分别表达的，但实际上是同样的事物的正反两个方面。

4. Therefore it is important for a contractor to ensure that，where the main contract contains a unilateral termination clause，the sub-contract contains a similar provision.

译文：因此，重要的是，承包商应确保当主合同包含单方终止合同的条款时，分包合同也

应该包含类似的条款。

5. Whether or not the employer's obligation extends to the obtaining of any official permits for the proposed work, such as planning permission, depends essentially on the terms of the contract.

译文：业主义务是否包含为准备实施的工程项目办理官方许可证，如规划许可证，主要取决于合同条款。

6. Indeed, where the contract envisages the use of an architect or an engineer to supervise the works and act as the employer's agent, the appointment thereof should be a condition precedent to the performance of the contractor's obligations.

译文：的确，如果合同指派建筑师或工程师来监督工程并担任业主代表，这项任命应该是承包商履行义务的先决条件。

Notes：中的 there 代表 an architect or an engineer to supervise the works and act as the employer's agent 以求句子的简洁。因此 the appointment thereof 可译成“这项任命”、“此任命”或“该项任命”等。

7. However, any responsibility for the defaults of nominated sub-contractors or suppliers must be based on an express term of the contract, namely it is not necessarily imposed on the part of employer.

译文：不过，指定分包商或供货商的任何违约责任都必须以合同的明示条款为依据，即：这些责任并不一定都被视为业主的责任。

8. Thus, while it is the contractor's right to complete the project on or before the contractual completion date, the employer's duty of co-operation extends only to ensuring that the completion date is achieved.

译文：这样，当按时或提前竣工是承包商的权利时，业主的合作职责仅限于确保按时竣工。

7.2 Express terms of Employer's Obligations

In addition to the implied obligations described above, the express terms of any construction contract will undoubtedly impose duties on the employer. Some of these (such as the duty to give possession of the site) will go no further than the term which would in any event be implied. Others, such as duties to insure, permits, licences, or approvals. Naturally, the range and content of the employer's express obligations will vary from one form of contract to another, and it would not be practicable here to deal with all the possibilities. We shall therefore look at those obligations contained in FIDIC Conditions of Contract, in order to give some idea of what is commonly found.

7.2.1 Payment

Undoubtedly the most important of all the employer's express obligations is to pay the contractor the sum of money which forms the consideration for the contract, known as the contract sum. Besides, FIDIC Conditions of Contract provides that the employer shall make an advance payment, as an interest-free loan for mobilization, under the condition that the contractor submit a guarantee against it.

7.2.2 Necessary Nominations

Under FIDIC Conditions of Contract, the employer shall appoint the Engineer as a contract administrator, who shall carry out the duties assigned to him in the contract. However, the employer shall not replace the Engineer with a person against whom the contractor raises reasonable objection by notice to the employer, with supporting particulars.

7.2.3 Site Obligations

The need to give the contractor possession of the site at the right time is fundamental to the contract. Failure to do so will cause the employer to forfeit any claim for liquidated damages on late completion. It will also render the employer liable to pay damages to the contractor. Clause 2.1 of FIDIC Conditions of Contract requires the employer to give the contractor right of access to, and possession of, all parts of the site within the time stated in the Appendix to Tender. A failure to secure possession for the contractor will constitute a breach by the employer.

Unless otherwise agreed, the employer's obligation is to give possession of the whole of the site from the outset, and not just those parts where work is to begin immediately. It is further accepted that the contractor must be given not only the actual area to be built on, but also sufficient surrounding space to enable the work to be properly undertaken. This would include, for example, room to erect temporary buildings and compounds, to store equipment and so on. However, the employer will only be obliged to provide access to the site across adjoining land where this is specifically stated in the bills or contract.

7.2.4 Insurance

Clause 18 of FIDIC Conditions of Contract offers a choice of insurance arrangements, depending upon whoever is designated the insuring party for their own respective insurance responsibilities which are usually specified in the relevant sub-clauses or appendix to the tender.

7.3 Responsibility for the Contract Administrator

It has been emphasized above that many construction contracts allocate to the contract administrator the performance of what are really the employer's obligations. These will be considered in more detail; our interest for the moment is in the extent to which the employer can be held responsible to the contractor for the contract administrator's actions.

It is crucial in this context to distinguish between two separate aspects of the contract administrator's role. First, there are the numerous things which are done in the capacity of agent of the employer (such as supplying information, giving instructions as and when necessary, and so on). In many proceedings, it was held that the employer impliedly guarantees that these functions will all be performed 'with reasonable diligence and with reasonable skill and care'. This means that any negligence on the part of the contract administrator will render the employer liable for breach of contract to the contractor. The employer will then of course be able to claim against the contract administrator for breach of the terms of engagement.

The second aspect of the contract administrator's role concerns certain 'independent' or 'discretionary' functions, such as adjudication or certification. As to these, all the express and implied terms of contract as well as the concerning laws provide that the employer ensure that the contract administrator will exercise his discretionary powers reasonably; and although the contract administrator may be engaged or employed by employer, he will leave him free to exercise his discretion fairly and without improper interference by him. As a result, an employer who obstructs or interferes with the issue of a certificate, or who puts improper pressure on the contract administrator, will be liable to the contractor.

In spite of this statement, in some circumstances the employer's duties in respect of the contract administrator's 'discretionary powers' go further than mere non-interference. For example, when an employer became aware that the contract administrator was acting improperly, by refusing to deal at all with applications for extension of time, it is the employer's obligation in such circumstances to order the contract administrator to carry out his duties under the contract and, if that failed, to dismiss and replace him.

Words, Phrases and Expressions

licence *n.* 许可证，执照（permit）
consideration *n.* 报酬，补偿费（reward）
advance payment 预付款（down payment）
mobilization *n.* 动员
guarantee *vt.* 保证（to ensure）
possession *n.* 占有，拥有
forfeit *vt.* 丧失，失去（to lose）

render *vt.* 给予（to administer）
compound *n.*（筑有围墙的）院子
adjoining *adj.* 邻接的，相邻的
insuring party 投保方
crucial *adj.* 决定性的（decisive）
in the capacity of 以…的身份，以…的资格
proceeding *n.* 诉讼（lawsuit）
function *n.* 职责（duty）
adjudication *n.* 裁决，判决（judge）
interference *n.* 干涉，干扰
obstruct *vt.* 阻碍（to hinder）

疑 难 词 句

1. Some of these (such as the duty to give possession of the site) will go no further than the term which would in any event be implied.

译文：这些义务中的某些义务（如给予工地的占有权的职责）将不会超出任何情况下的默认条款。

2. It is further accepted that the contractor must be given not only the actual area to be built on, but also sufficient surrounding space to enable the work to be properly undertaken.

译文：进一步公认的是：承包商不仅应获得该建筑物之下的实际区域的占有权，而且还应获得其周围的足够空间，以便能适当地履行该项目。

3. However, the employer will only be obliged to provide access to the site across adjoining land where this is specifically stated in the bills or contract.

译文：但是，业主只有义务向承包商提供穿越工程量清单或合同专门指定的相邻地域到达工地的通道的使用权。

4. Our interest for the moment is in the extent to which the employer can be held responsible to the contractor for the contract administrator's actions.

译文：我们现在想要了解的是业主能为合同管理者的行为向承包商承担的责任范围（限度）。

5. And although the contract administrator may be engaged or employed by employer, he will leave him free to exercise his discretion fairly and without improper interference by him.

译文：虽然合同管理者可能受雇于业主，但业主会给予合同管理者公正地自主行使权力的自由，而这种权力的行使将不会受到业主的不当干预。

6. In spite of this statement, in some circumstances the employer's duties in respect of the contract administrator's 'discretionary powers' go further than mere non-interference.

译文：尽管如此，在某些情况下业主对于合同管理者的“自主权力”不会仅限于不干预。

7.4 Responsibility for site conditions

A common question which has arisen very often over the years concerns the extent to which an employer bears legal responsibility in respect of the feasibility of the site conditions. In particular, if site conditions such as the underground condition turn out to be unexpectedly adverse, which would make the proposed method of working prove impracticable, is there any way in which the contractor can claim compensation? The answer to these questions naturally lies within the terms of the contract, but a fairly clear overall picture emerges from the experience of many occurrences over years. This general picture is discussed below to establish an international practice in this respect.

7.4.1 Contractor's Risk

As a basic principle, it is clear that the risk of adverse site conditions rests with the contractor. This was firmly established in many cases of underground works. Taking a project of the carrying out of sewerage works as an example: because the soil was softer than anticipated, the contractor had to carry out considerable extra work. When the engineer refused to authorize extra payment for this work as a variation, the contractor left the site and claimed a reasonable sum for the work done. No boreholes had been sunk in advance by either party, but the employer had received reports before signing the contract that the contractor was certain to lose money in the type of ground to be expected. It was nonetheless held that the employer owed no duty to disclose these reports to the contractor, whose claim accordingly failed.

The principle that adverse site conditions are a contractor's risk is not altered merely because the employer provides plans or specifications at the time of tender. The mere fact that these are provided does not imply any warranty by the employer as to their accuracy. Therefore, the contractor who accepts such a plan should examine and certify its reasonableness and feasibility with a professional way. In a railway construction project, for example, a contractor undertook to build a railway in Brazil for a lump sum. The engineer's plans proved to be hopelessly inadequate and, as a result, the contractor was forced to excavate about twice as much as had been anticipated. It was held that, since the accuracy of the plans was in no way warranted, the contractor was not entitled to any extra payment in respect of this work.

As with site conditions, so with working methods. Many cases turned out to be a risk on contractors caused by a wrong working method. For example, in a contract for the demolition and replacement of a bridge. The plans and specifications prepared by the engineer (whose directions the contractor was required to obey) featured the use of caissons to enable work to be done whatever the state of the tide. Unfortunately, these caissons proved to be useless, with the result that the contractor suffered considerable delay and extra expense. Once again, however, it was held that there was no implied warranty that the bridge could be built in the manner specified, and so the risk lay on the contractor.

7.4.2 Employer's Responsibility

Notwithstanding the general principle outlined above, there may be situations in which an employer is liable when the project proves to be unexpectedly difficult or expensive to carry out. The major ways in which this occurs are as follows:

- *Implied warranty*. Although an employer is not obliged to guarantee the accuracy of the tendering information he provides, it may occasionally be possible for a warranty of accuracy to be implied. In a project, for example, a contractor tendered to design and build six blocks of dwelling houses, under instructions to design the foundations on certain hypotheses as to ground conditions. These hypotheses, which were based on borehole data, subsequently proved inaccurate. It was held that, in these circumstances, the employer must be taken to have warranted that the ground conditions would be as they were hypothesized to be.
- *Misrepresentation*. Although there is no warranty as to site conditions, a contractor may be able to show some misrepresentation given by or on behalf of the employer. This will certainly be the case if there is any deliberate fraud in covering up the true conditions. Even in the absence of fraud, it is no doubt possible to base a claim on the employer's negligent misrepresentation. However, this will not be easy to establish, given the common contractual provision that it is the contractor's responsibility to check such matters anyway.
- *Standard method of measurement*. Building contracts frequently state that the bills of quantities on which they are based have been prepared in accordance with a specified Standard Method of Measurement. Where this is so, a contractor who unexpectedly encounters rock or some ground conditions which are not within the scope of a certain Standard may claim that this should have been a separate item in the bills and that the employer must therefore pay the extra cost. Although such claims may have been described as without foundation and without merit, there is some legal authority to suggest that they are valid.

Words, Phrases and Expressions

feasibility *n.* 可行性，可能性
impracticable *adj.* 不可行的（unworkable）
emerge *vi.* 出现（appear）
practice *n.* 惯例（tradition）
borehole *n.* 钻孔（drill-hole）

disclose *vt.* 透露（to reveal）
warranty *n.* 保证（guarantee）
lump sum 总价包干
demolition *n.* 拆毁（removal）
caisson *n.* 沉箱
tide *n.* 潮水
hypotheses *n.* 假定（assumption）
deliberate *adj.* 故意的，蓄意的（intended）
fraud *n.* 欺骗（cheat）
merit *n.* 价值（worth）

疑 难 词 句

1. A common question which has arisen very often over the years concerns the extent to which an employer bears legal responsibility in respect of the feasibility of the site conditions.

译文：多年来人们经常提到的一个普遍性的问题是关于业主在现场条件方面应承担的法律责任的限度问题。

2. Although an employer is not obliged to guarantee the accuracy of the tendering information he provides, it may occasionally be possible for a warranty of accuracy to be implied.

译文：虽然业主没有义务保证他提供的投标资料的准确性，但在某些情况下这种准确性的担保可能是默认的。

Notes: Under some circumstances, it is possible for an employer to be liable for the contractor's errors caused by the inaccuracy of the information provided by the employer.

3. However, this will not be easy to establish, given the common contractual provision that it is the contractor's responsibility to check such matters anyway.

译文：不过，这种索赔并不容易被认可，因为常见的合同条款都规定：承包商对业主提供的资料有核实的责任。

4. Although such claims may have been described as without foundation and without merit, there is some legal authority to suggest that they are valid.

译文：虽然这些索赔可能会被认为没有依据而且没有意义，但某些法律观点认为它们是具备法律效力的。

Chapter 8 Time

Time is an extremely important issue in construction. Together with cost and quality, it is a primary objective of project management, and a major criterion by which the success of a project is judged. The definition of time in a construction project can be summarized in a sentence: *On the Date of Possession possession of the site shall be given to the contractor who shall thereupon begin the Works, regularly and diligently proceed with the same and shall complete the same on or before the Completion Date.* This sentence identifies the three basic time-related issues as commencement, progress and completion. In fact there are also two other issues: the contractor's continuing obligations after completion, and the extensions of time which may be available to the contractor when the work is delayed by certain specified causes. Here, these issues are discussed as follows:

8.1 Commencement

The issues at the beginning of the contract involve giving possession of the site to the contractor, the timing of this possession and potential delays to the possession. Normally, possession should take place not more than two months after the successful contractor has been awarded the contract. Too speedy a start may, however, not necessarily be good for both parties by causing extra work and delay, rather than hastening the construction period. This needs to be balanced against the needs of the client to avoid undue delay which may cause extra costs.

8.1.1 Possession of the site

An employer who fails to give the contractor possession of the site may be liable to pay damages for breach of contract. However, the employer is not deemed to guarantee possession and will therefore not be liable if the contractor is prevented from gaining access by some third party, such as unlawful strike, over whose activities the employer has no control.

Some standard form contracts entitle the contractor to possession of the whole site, even though access to some parts may not be required until a later stage of the project. By contrast, some other contracts only requires the employer to give possession of so much of the site as is required to enable the contractor to commence the works in accordance with the programme. In either case, however, what is given to the contractor must include not only the actual area to be built on, but also enough of the surrounding area to enable the work to be undertaken.

8.1.2 Date for possession

Most building contracts will specify a date on which the contractor is to be given possession of the site, after which the contractor may commence the works. If possession is not then given on the date specified, the employer will lose the right to recover liquidated damages from the contractor in the event of late completion. Some contracts offer the employer a little extra flexibility by providing that, if no date is specified from the outset, it is for the engineer to notify the contractor of the date for commencement of the works. This notice must be given in writing, and the date itself must be within 28 days of the award of the contract.

If the contract contains no specific commencement provision, then the contractor must be given possession at such a time as will enable the work to be completed by the completion date. The contractor is not obliged to start work on the date for possession, as such; however, a contractor who does not start reasonably quickly may be liable for not proceeding 'regularly and diligently' or 'with due expedition and without delay'.

8.1.3 Deferred Possession

Although the engineers' authority to order the postponement of any work cannot be used by the employer so as to justify delay in giving the contractor possession of the site, there are specific provisions in some standard form contracts under which the employer may defer the date for possession by up to six weeks. In order for this provision to apply, it must be stated in the Appendix to the contract.

8.2 Progress

Usually, most contracts only specify a fixed date for completion, but makes no provision as to the rate at which the works are to progress, therefore the contractor has absolute discretion as to how the work is planned and performed, provided only that it is completed on time. Furthermore, while many contracts require the contractor to submit a programme for the execution of the works, this in itself does not mean that there is a contractual obligation to keep to that programme. Indeed, it should be appreciated that, if there were such an obligation as to the rate of progress, it would apply to both parties. Thus the employer would have to ensure that the contractor was provided with all necessary information at such a time as to enable compliance with the programme.

From an employer's point of view, it would be very inconvenient to have no control at all over the progress of the contract works. It is for this reason that most construction contracts require the contractor to maintain a satisfactory rate of progress throughout the project by stating that the works should be executed 'regularly and diligently' or 'with due expedition and without delay' and also many contracts make the failure to do so a ground on which the employer can determine the contractor's employment.

By 'regularly and diligently', the word 'regularly' means at least that the contractor must attend on a daily basis with sufficient labour and materials to progress the works substantially in accordance with the contract. 'Diligently' refers to the need for the contractor to apply that physical capacity industriously and efficiently. Taken together, the contractor's obligation is 'to proceed continuously, industriously and efficiently with appropriate physical resources so as to progress the works steadily towards completion substantially in accordance with the contractual requirements as to time, sequence and quality of work'.

Words, Phrases and Expressions

criterion *n.* 标准（standard）
deem *vt.* 视为（to treat）
commencement *n.* 开工（start，beginning）
continuing obligation 后续义务（following obligation）
hasten *vt.* 加速（to quicken）
postponement *n.* 延期，推迟（delay）
justify *vt.* 证明…合理
keep to 遵守（to obey）
ground *n.* 理由（reason）
employment *n.* 雇佣（engagement）
attend *vt.* 照料，照顾（to look after）
substantially *adv.* 充分地（fully）
physical *adj.* 物质的
industriously *adv.* 勤劳地，勤奋地（diligently）
steadily *adv.* 稳定地（smoothly）
sequence *n.* 次序，顺序（order）

疑 难 词 句

1. Too speedy a start may, however, not necessarily be good for both parties by causing extra work and delay, rather than hastening the construction period.

译文：但是，开工太早对双方都不一定是好事，因为这样会引起额外的工作和延误，并不会缩短建设工期。

2. Some standard form contracts entitle the contractor to possession of the whole site, even though access to some parts may not be required until a later stage of the project.

译文：有些标准格式合同给予承包商拥有全部工地的权利，尽管工地的某些部分在以后的施工阶段才需要进入。

3. Taken together, the contractor's obligation is 'to proceed continuously, industriously and efficiently with appropriate physical resources so as to progress the works steadily towards completion substantially in accordance with the contractual requirements as to time, sequence and quality of work'.

译文：这两个词放在一起的意思是：承包商的义务是不间断地、勤奋地、高效地利用适当的物质资源充分地根据合同关于时间、顺序和质量的要求稳定地推进项目的进展直到竣工。

8.3 Completion

In building contracts, completion is a vague concept. The fact that building projects can be handed over in a less than perfect state is to the advantage of both parties.

8.3.1 Meaning of Completion

A contractor cannot truly be said to have totally performed the contract if a single item of work is missing or defective. From a practical point of view, however, to delay the handover of something as complex as a large building for a trivial breach would cause enormous inconvenience. As a result, most building contracts require the contractor to bring the works to a state described by such expressions as practical completion or substantial completion.

Whether or not a building is 'complete' in this sense is normally a decision of the contract administrator, based on an inspection of the works and the exercise of reason and judgement. As to precisely what is required, the international practice held by different courts have ranged between two extremes. In some cases, the Court of Appeal adopted a very functional (and liberal) approach. He defined practical completion as 'completion for all practical purposes, that is to say for the purpose of allowing the employers to take possession of the works and use them as intended'. However, as for other cases, the court may take the much stricter view that 'what is meant is the completion of all the construction work that has to be done'.

Many in the construction industry would undoubtedly favour the first of these opinions as being the more practical. It is also worthy of note that the only standard form contract which seeks to define practical completion adopts this approach.

However, the current judicial view is much closer to that of later opinion. It appears that a contract administrator may certify practical completion notwithstanding that there are trivial defects or omissions in the works, but should not do so where there are serious defects which go beyond the merely trivial.

8.3.2 Date for Completion

It is not essential for a building contract to specify a date for completion; if it does not, the contractor's implied obligation is to complete the works within a reasonable time. To rely on such an obligation, however, is not very satisfactory, at least on a project of any appreciable size, for the client will normally need to have some degree of confidence about precisely when to expect completion. In addition, without a specific completion date there can be no provision for 'liquidated damages', that is a fixed sum to be paid by the contractor for every day or week of delay.

It is thus usual to name the date by which completion is required. In many standard form contracts, this is done by means of an entry in the Appendix to the contract. Moreover, even where no precise date has been included in the contract itself, a court may be persuaded to imply a term for completion by a certain date, on the ground that the parties must have intended this.

8.3.3 Delay in completion

The contractor's obligation to complete the works by the completion date is, like all such obligations, backed up by legal sanctions. Under certain types of contract (for example contracts for the sale of perishable goods), time is expressly or impliedly 'of the essence'. Where this is so, any lateness in performance entitled the other party to determine the contract. However, construction contracts very rarely fall into this category. Consequently, the employer's remedy for late completion will be an award of damages for breach of contract.

As to how such damages are to be measured, it is of course perfectly possible for the contract to say nothing, and to leave the assessment of the employer's loss (including any loss of profit) to an arbitrator or a court. However, it is standard practice in building and civil engineering contracts to state in advance what the damages shall be for delay, and this is usually done by specifying a fixed sum of money to be due for every day, week or month by which the contractor fails to meet the prescribed completion date. Such sums are called liquidated damages, liquidated and ascertained damages.

8.3.4 Sectional completion and partial possession

If the intention of the parties is that the contract work should be completed and handed over in phases, it is essential that the contract documents make proper provision for this. Many standard form contracts, for instance, enable different Sections to be identified in the Appendix, each with its own time for completion and its own liquidated damages.

If there is no provision in the contract for the works to be completed and handed over in sections, the contractor's right to occupy the whole of the site will continue as long as the work under the contract is not completed. However, some contracts enable this right to be given up by the contractor. Usually, if the employer requests it and the contractor agrees (and there must not be an unreasonable refusal), the employer can take possession of those parts of the works which are ready. Where this is done, the contract administrator must issue a written statement identifying the parts taken over and, for most purposes, practical completion of those parts is deemed to have occurred. It should also be noted that in this case the employer may be permitted by the contractor to use parts of the site for 'storage of his goods or otherwise', provided that this will not prejudice the insurance position. Once again contractor's consent is not to be unreasonably withheld.

8.3.5 Effects of completion

When the contract administrator certifies that the works have been completed, a number of consequences will follow. Precisely what these are will depend upon the terms of the contract concerned, but the following are typical:

- The employer is entitled and obliged to take possession of the contract works.
- The contractor's responsibility (if any) for insuring the contract works comes to an end.
- Any liability of the contractor to pay damages for late completion ceases. Moreover, this liability will not be revived if the work is later found to contain defects, for such a discovery does not retrospectively invalidate the certificate.
- The contractor usually becomes entitled to the release of one-half of the accumulated retention money.
- The Defects Liability Period begins.

8.4 Contractor's Obligations after Completion

There are further obligations which are imposed on the contractor after completion. Under the most standard form contracts, the issue of the 'Certificate of Practical Completion' marks the start of the 'Defects Liability Period', which usually lasts one year in the Appendix. Any defects, shrinkages or other faults which arise during this period due to defective materials or workmanship must be put right by the contractor at his own expense. Further, the contractor is responsible for damage which appears during this period, but only where the contract administrator certifies that the damage was actually caused before practical completion.

The contract administrator must issue a schedule of such defects to the contractor not later than 14 days after the end of the defects liability period, and the contractor then has a reasonable time to put them right. Once this has been done, the contract administrator will issue a 'Certificate of Completion of Making Good Defects', following which the contractor becomes entitled to the remaining part of the retention money. If the contractor does not make good the defects, after being served with a schedule of defects, then the employer can give notice that the work must be done. If the contractor does not comply with that notice immediately, then the employer can employ others to do the work, and recover the expense from the contractor.

Furthermore, the contractor's obligation may go beyond repairing and completing those items which are a direct consequence of the contractor's default. However, work done on other items (i. e. those which are not due to workmanship or materials being in breach of contract) entitles the contractor to extra payment.

It is sometimes said that, during a defects liability period, the contractor has the right as well as the obligation to put right any defects which appear. What this means is that an employer who discovers defects should operate the contractual defects liability procedure, rather than either appointing another contractor to carry out the repairs or simply suing the contractor for breach of contract. Actually, this can be done as a result of the contractor's failure to fulfill the contractual defects liability procedure.

Words, Phrases and Expressions

vague *adj.* 模糊的，不明确的（ambiguous）
hand over *vt.* 移交，交出（to submit）
trivial *adj.* 微不足道的，不重要的（insignificant）
liberal *adj.* 宽容的，慷慨的（not strict）
favour *vt.* 赞同（to agree）
judicial *adj.* 司法的（legal）
appreciable *adj.* 可观的，大的（substantial）
legal sanction 法律制裁
determine *vt.* 终止（to terminate）
assessment *n.* 评价（evaluation）
phase *n.* 阶段（stage, section）
prejudice *vt.* 不利于，损害（to harm）
consequence *n.* 后果（aftereffect）
revive *vt.* 恢复（to resume）

retrospectively *adv.* 回顾地，追溯地（backward）
invalidate *vt.* 使作废，使无效
shrinkage *n.* 收缩
make good *vt.* 修复（to repair）

疑 难 词 句

1. It appears that a contract administrator may certify practical completion notwithstanding that there are trivial defects or omissions in the works, but should not do so where there are serious defects which go beyond the merely trivial.

译文：看来，合同管理者可以在工程存在一些小的瑕疵或遗漏的情况下证明该工程已基本完工，不过当工程存在严重瑕疵而不仅仅是小瑕疵时，合同管理者不能这样做。

2. Consequently, the employer's remedy for late completion will be an award of damages for breach of contract.

译文：因此，业主对推迟竣工的补救办法将是获得一些违约损害的赔偿。

Notes：damages 指的是 liquidated damages 即：误工损害赔偿金

3. Moreover, this liability will not be revived if the work is later found to contain defects, for such a discovery does not retrospectively invalidate the certificate.

译文：此外，该责任将不会因以后工程被发现有瑕疵而重新被追究，因为这种发现不能追溯性地废除竣工证书。

4. Further, the contractor is responsible for damage which appears during this period, but only where the contract administrator certifies that the damage was actually caused before practical completion.

译文：此外，承包商应对该期间内工程的任何毁坏负责，但前提是合同管理者能证明该毁坏确实是基本竣工前导致的。

8.5 Extensions of Time

Most building contracts contain express provisions under which the period allowed for the contractor to undertake and complete the works can be extended. These provisions apply to delays which are neither the fault nor the responsibility of the contractor. Such provisions obviously benefit the contractor, who will not be liable to pay damages for delay during the period for which time is validly extended. In addition, and less obviously, the power to extend time is also for the employer's benefit, for the following reason.

At common law, the contractor's obligation to complete the works by the specified date is removed if the employer delays the contractor in the execution of the works. Therefore, if the contract administrator issues an instruction which increases the amount of work to be done, or is late giving the contractor necessary information or instructions, the specified completion date no longer applies. In this situation, the original time limit is invalid, and the contractor's obligation is merely to complete the works within a reasonable time. In order to fix what is 'reasonable', all the circumstances of the particular project must be taken into account, but in many cases it will simply mean that the amount of delay for which the employer is responsible will be added to the old completion date.

The importance of invalidating the fixed date is that a contractor who has caused part of the delay is still liable to pay general damages for delay, but is not liable for liquidated damages. Even where the delay caused by the employer is a very small part of the overall delay, the employer cannot simply discount this and claim liquidated damages for the remainder. In this case, the liquidated damages provision fails altogether, and the employer can claim only for those losses resulting from the delay which can actually be proved. Such proof is sometimes difficult, and in any event the amount recovered may be less than what was fixed as liquidated damages. An employer who has caused delay therefore has a very strong interest in being able to extend time for this, so as to retain the entitlement to liquidated damages from the revised completion date.

8.5.1 Grounds for extensions of time

A fundamental point is that the time for completion can only be extended where the contract permits, and strictly in accordance with the contract provisions. If delay is caused by some event which the contract does not cover, then the contractor cannot claim an extension, nor can the employer insist on giving one (in order to keep alive a claim for liquidated damages). For example, it was held that a power to extend time for delays caused by the ordering of extra work only applied where the extra work was properly ordered, and not where the architect gave the relevant instructions orally instead of in writing, as the contract required.

8.5.2 'Relevant events' as a ground for extensions of time

As a good example of the kind of grounds on which time for completion can be extended in building contracts, we can take a list of 'relevant events' as follows:

Force majeure

The phrase 'force majeure' derives from French law where it is used to describe situations where an unforeseeable event makes execution of the contract wholly impossible and is of such intensity that it cannot be overcome. Its inclusion as a ground for extending the contract period or terminating the contract is usually in the express terms of all the contracts.

Exceptionally adverse weather conditions

Exceptionally adverse weather is not restricted to bad weather. Since excessively hot and dry weather can also cause problems with progress of the works, this phrase was introduced in some contracts to include all exceptional weather conditions, rather than merely cold and wet weather.

However, while other contracts use the same phrase, they do not recognize any kind of weather conditions as grounds for extensions of time. This means that, under some other forms of contract, the entire 'weather risk' is borne by the contractor (who will presumably price the tender accordingly). Under some forms of contract, the risk is shared.

In some countries, it can be difficult to establish exactly what constitutes 'exceptionally adverse weather'. An examination of local weather records should establish what is normal for the locality. This will provide a definition of what is 'usually adverse' weather and thus help to identify 'exceptionally adverse' weather. However, since the exact nature of the weather is critical in deciding the validity of a claim, Meteorological Office records may be inadequate. It is wise to keep detailed site weather records because weather can vary greatly over short distances, especially where there are hills nearby.

In considering weather problems, it must always be borne in mind that there can be no extension of time unless the whole project is actually or potentially delayed. If, for example, exceptionally adverse weather occurs at a time when most of the work is indoors, this is not a ground for an extension. It is also important to note that it is the actual effect of weather on the work which is relevant, thus what matters is the weather at the time when a particular part of the work was actually carried out, not necessarily at the time when it was programmed to be carried out.

Loss or damage occasioned by the Specified Perils

The 'Specified Perils' are listed in most contracts and include such matters as fire, lightning, explosion, storm, flood, bursting or overflowing of water tanks, apparatus or pipes etc, riot and civil commotion. Most contracts allow for an extension of the contract period if loss or damage caused by these things delays the contractor. These perils are the subject of the insurance provisions in most building contracts.

Civil commotion, strike, lock-out etc.

This applies to delays caused by industrial action taken by any group of people who are employed on the works, are manufacturing items for inclusion in the works or are involved in transportation of goods to the works.

Compliance with the contract administrator's instructions

Most contracts give the contract administrator wide powers to issue instructions to the contractor during the progress of the works. In relation to some, though not all, of these, most standard form contracts provide that any delay resulting from the contractor's compliance will bring an entitlement to an extension of time. Instructions on the following matters fall into this category:

- discrepancy in or divergence between contract documents;
- variations;
- the expenditure of provisional sums;
- the postponement of any work to be executed under the contract;
- any action to be taken concerning fossils, antiquities and other objects of interest or value;
- nominated sub-contractors;
- nominated suppliers.

It is important to appreciate that, the contractor must comply with all instructions which the contract empowers the contract administrator to issue. If the contractor fails to do so, the employer may employ others to do part of the works. In these circumstances the contractor would not be entitled to claim an extension of time.

Opening up and inspection of defective work

Some contracts cover a further issue which may result from an instruction by the contract administrator, namely, the opening up and testing of work which has been covered up. Delay resulting from this will entitle the contractor to an extension of time, unless the inspection shows the work to be defective.

Delay in the supply of information

It is expected that the contract administrator will keep the contractor supplied with the information needed to carry out the work. Failure to supply drawings or details or information on levels at the right time will be a breach of contract for which the contractor will be entitled to recover damages. Moreover, it is the contract administrator's responsibility to know when information is going to be required and to ensure that it is ready.

Furthermore, as well as leading to an action for damages, delay in supplying necessary information may in certain circumstances be a ground on which the contractor is entitled to an extension of time. For this purpose, 'information' is defined as 'necessary instructions, drawings, details or levels', and the 'instructions' referred to cover only those matters about which the contract gives the contract administrator power to issue instructions.

Unlike the claim for damages, above, in order for a contractor to claim an extension of time on the grounds of late supply of information, the information must have previously been requested by the contractor in writing, at a time neither unreasonably distant from nor unreasonably close to the time when it is required. It is worthy to note that the submission of the contractor's original programme to the contract administrator should come at a early date as required in the contract. In any event, the insertion of due dates for information, and the regular revision and re-issue of such a programme, will certainly fulfil the requirements of this clause.

Delay on the part of nominated sub-contractors/suppliers

Many building contracts have a clause as to this subject with such practical implications that if the main contractor is entitled to an extension of time when delayed by a nominated sub-contractor, there is no liability on the part of the contractor for liquidated damages. This in turn means that, although the contractor may claim damages from the nominated sub-contractor for loss caused by disturbance of the progress of the works, these damages will not include the amount of the liquidated damages, because the contractor will not have paid these. As a result, unless the employer has a direct claim against the nominated sub-contractor for the delay, the employer will simply lose out and the nominated sub-contractor will evade responsibility.

The contractor should closely monitor the progress of nominated sub-contractors to ensure that they remain on programme, and so that the contractor can warn the contract administrator in plenty of time of the likelihood of a nominated sub-contractor delaying the works. Similarly, supplies of goods and materials from nominated suppliers should be closely checked. Otherwise, contractors may lose their right to an extension of time, either because they have not taken all practicable steps to avoid or reduce the delay or, more generally, because they have not constantly used their best endeavours to prevent delay.

It is important to note that, if a nominated sub-contractor ceases work altogether (for example because of insolvency), this is not in itself grounds for the main contractor to claim an extension of time. The position is that the contract administrator has a duty to renominate another sub-contractor within a reasonable time, and that it is only delay in doing so that will entitle the main contractor to extra time.

The execution of work not forming part of the contract

Many standard form contracts allow the employer to undertake parts of the works, either directly or through another contractor, where this is stated in the contract. Such work will not form part of the contract, but a delay in that part of the work will give the contractor grounds for extension of time.

The supply of materials by the employer

Besides, the employer may choose to supply some of the materials or goods for incorporation in the works. Once again, however, this must be clearly stated in the contract since, if this is not done, it is the contractor's obligation, and right, to do the entire work set out in the contract documents.

Where the employer is to supply materials, the contractor must make sure that the employer is fully aware of the dates upon which they will be required. If this is not done, the contractor will lose the right to claim an extension of time for any resulting delay.

The exercise by the government of any power which directly affects the works

Any affect on the execution of the works caused by exercise of the governmental decrees or regulations may entitle the contractor to an extension of time or extra payment by the employer. An example of the kind of exercise of government power which would be relevant here is the three-day week which resulted from the miner's strike in 1974. In an effort to break the miners' strike, the government decreed that people could only have electricity for three days per week, with a view to preserving coal stocks for as long as possible.

In order for this clause to be used as a reason for claiming an extension of time, the contractor must have had no knowledge of the matter at the date of tender.

Contractor's inability to secure labour, goods or materials

Some contracts allow for an extension of time or adjustment of the contract price as a result of shortages or market price fluctuation of the labor, goods, or materials needed by execution of the works. However, It must be very difficult to establish the extent to which shortages of materials exist - particularly if paying a higher price than the market rate would secure them. Clearly, the price paid must bear some relationship to what was envisaged at the time of tender, but if the contractor waits for the shortage to finish, the price may never drop. In order to be able to claim an extension of time or extra payment under this provision, it must be demonstrable that the contractor could not have reasonably foreseen such difficulties at the date of tender.

Deferment of Date of Possession

As we noted before, failure by the employer to give the contractor possession of the site on the agreed date is a serious breach of contract. However, the employer may defer the giving of possession for whatever period (not exceeding six weeks in many contracts) is specified in the appendix. If possession is deferred in this way, then naturally the contractor is entitled to an extension of time.

Words, Phrases and Expressions

extend *vt.* 延长（to postpone, to prolong）

power *n.* 权力（authority）

overall *adj.* 全面的，总体的（whole）

remainder *n.* 剩余部分（rest）

derive from 由…起源

unforeseeable *adj.* 不可预测的（unpredictable）

recognize *vt.* 承认（to accept）

presumably *adv.* 大概地，可能地（probably）

potentially *adv.* 潜在地

occasion *vt.* 引起（to cause）

peril *n.* 危险（danger）
entitlement *n.* 权利，资格（right，title）
empower *vt.* 授权（to authorize）
level *n.* 标高，水平高度（elevation）
due *adj.* 到期的（mature）
disturbance *n.* 打扰，扰乱（upset）
evade *vt.* 逃避（to avoid）
likelihood *n.* 可能性（probability）
constantly *adv.* 经常地，不变地（often，frequently）
endeavour *n.* 努力（effort）
renominate *vt.* 再提名，再任命（to reappoint）
set out *vt.* 安排（to arrange）
decree *n.* 法令，政令（statute）
shortage *n.* 短缺（lack）
demonstrable *adj.* 可论证的，可证明的（provable）

疑 难 词 句

1. The importance of invalidating the fixed date is that a contractor who has caused part of the delay is still liable to pay general damages for delay，but is not liable for liquidated damages.

译文：废除固定的竣工日期的重要性在于：需要对延误承担部分责任的承包商仍然有责任赔偿延误造成的一般损害，但没有责任支付误期损害赔偿金。

2. An employer who has caused delay therefore has a very strong interest in being able to extend time for this，so as to retain the entitlement to liquidated damages from the revised completion date.

译文：造成延误的业主因此会有延长工期的强烈愿望，以便能够继续获得从修改后的竣工日开始计算的误期损害赔偿金的权利。

3. It is also important to note that it is the actual effect of weather on the work which is relevant，thus what matters is the weather at the time when a particular part of the work was actually carried out，not necessarily at the time when it was programmed to be carried out.

译文：同样值得特别注意的是：这种影响指的是天气对相关工程的实际影响。就是说，关键是实际实施该具体工程时的天气，而不一定是计划实施该工程时的天气。

4. Some contracts cover a further issue which may result from an instruction by the contract administrator，namely，the opening up and testing of work which has been covered up.

译文：有些合同涉及由合同管理者的指示引起的更进一步的问题，即重新将已经覆盖的工程挖开并进行检验而引起的相关问题。

5. Furthermore, as well as leading to an action for damages, delay in supplying necessary information may in certain circumstances be a ground on which the contractor is entitled to an extension of time.

译文：此外，除了能导致损害赔偿外，这种延期提供必要资料的行为在某些情况下也可以成为承包商有权延长工期的理由。

6. Unlike the claim for damages, above, in order for a contractor to claim an extension of time on the grounds of late supply of information, the information must have previously been requested by the contractor in writing, at a time neither unreasonably distant from nor unreasonably close to the time when it is required.

译文：与上述损害赔偿不同，为了使承包商能够以推迟提供资料为理由提出延长工期的索赔，承包商必须在此前曾经书面要求业主提供该资料，而该要求的提出时间既不能无理地（过分地）远离又不能无理地（过分地）接近合同规定提供该资料的时间。

7. Many building contracts have a clause as to this subject with such practical implications that if the main contractor is entitled to an extension of time when delayed by a nominated sub-contractor, there is no liability on the part of the contractor for liquidated damages.

译文：许多建筑合同都有包含以下含义的相关条款：如果总承包商有权因指定分包商的延误而延长工期，总承包商就因此没有支付误期损害赔偿金的责任了。

Chapter 9 Payment

The provisions relating to payment concern the way in which the contractor is paid by the employer. The contract sum which is given by the employer to the contractor is not always a fixed amount of money. However, there are only certain circumstances in which the contract sum can be altered. The most important of these is where there are variations, but there are others. Here we are going to discuss these issues, and also the way in which mechanisms such as retention and deduction affect payment to the contract.

9.1 Employer's obligations to pay

The primary obligation upon the employer is to give the contractor the sum of money which forms the consideration for the contract. Money must be paid promptly and fully unless there are specific reasons for withholding it.

9.1.1 Contract price

The contract price is dealt with in different ways by different contracts. Under some contracts, with bills of quantities, the bid by the contractor is based upon the work which is described and quantified in the contract bills. If any quantities are altered because of variations in the client's requirements, then the contract sum will be altered. Otherwise, the contractor is paid the amount of the tender. However, some other contracts, by contrast, are known as a 're-measurement', or 'measure and value' contracts. This means that even though there may be quantities in the bills which the contractor priced, their purpose was purely for the tendering process. The price to be paid for the works will not be fully established until the works are complete.

These two types of contract are basic types, and naturally there are variations on these basic subjects. The difference between them is of fundamental importance to any employer when considering which type of contract is appropriate for a particular project.

9.1.2 Time of payment

It would be rare for the employer to withhold all money until the contractor has fulfilled the contract. The employer is obliged by most contracts to pay to the contractor the contract sum by instalments. One of the main purposes of interim payments is to reduce the need for the contractor to fund the development of the project. This is because the total value of a each contract forms a large proportion of a contractor's annual turnover. Payment by instalments should eliminate the need for the contractor to borrow money until final payment.

The amount of money due in each instalment is recorded by the contract administrator in an 'Interim Certificate'. The issue of such a certificate by the contract administrator imposes upon the employer a strict obligation to make payment.

Some contracts oblige the employer to pay within 14 days the amount of money shown on an interim certificate. FIDIC Conditions requires the contractor to submit a monthly statement to the engineer. Within 28 days of the submission of this statement, the engineer shall certify, and the employer shall pay, the relevant amount of money.

In the event of the main contractor defaulting, there are usually provisions for the employer to make direct payments to sub-contractors.

9.1.3 Effect of certificates

Interim certificates exist simply as a mechanism for confirming that an instalment of the consideration is due to the contractor. Some contracts are based on a contract administrator's valuation of work done each month. FIDIC Conditions is based on a contractor's claim for payment which is evaluated by the engineer.

Whichever method is used to calculate the amount of money due, an interim certificate is not conclusive about anything. It says nothing about quality of materials or workmanship, nor does it indicate satisfaction with the work done to date. Anything which has been included in such a certificate may yet be the subject of a later certificate. It is only the final certificate which is ever conclusive. As a result, the only obligation which arises from an interim certificate is an obligation on the employer to make a payment within the stated time. Failure to do so is a serious breach of contract.

9.2 Contract sum

The price for the work is typically referred to as the contract sum or the contract price. It is important to know the specific definitions of what is included in the contract sum and how it can be changed.

9.2.1 Definitions in the contracts

In some contracts, the contract sum is defined in the Articles of Agreement as the amount of money which the employer will pay to the contractor. It is exclusive of VAT. It states that the contract sum is this amount of money or such other sum as shall become payable under the conditions, at the times and in the manner specified in the conditions. It is clear therefore that, under these contract terms, the sum is specified at the outset, although it may be altered as work proceeds.

In FIDIC Conditions, by contrast, the 'contract price' is defined as the sum to be ascertained and paid in accordance with the provisions. This concept of contract sum is very different from that in the contracts mentioned above. Under the FIDIC form, the sum can only be ascertained when the contract is concluded. The contract price the employer is finally obliged to pay is not a specified amount of money at the time the contract is signed.

In some contracts, the contract sum is fixed by a relevant clause, which may state that it cannot be altered or adjusted in any way other than by the conditions of contract. This also states that any errors in the computation of the contract sum, whether arithmetic or not, are deemed to be accepted by both parties. The effect of this clause is that the contract sum can only be changed when the conditions of contract allow adjustments.

9.2.2 Permissible changes

The contract sum may be changed for a variety of reasons, which can be divided into three groups as follows:

1. Reimbursement of the contractor for certain expense caused by the contract administrator, employer, or certain events outside the control of the contractor. These matters are covered by clauses of many contracts under which the contractor can claim for 'loss and/or expense'.

2. Payment for extra work brought about by an instruction of the contract administrator.

3. Reimbursement of extra expense brought about by market fluctuations affecting the contractor's inputs.

Words, Phrases and Expressions

contract sum 合同额
retention *n.* 保留，扣留（retaining）
deduction *n.* 扣除，减除（decrease）
promptly *adv.* 立即（immediately）
withhold *vt.* 扣留，保留（retain）
re-measurement *n.* 重新测量
measure and value 测量并估价
establish *vt.* 确定（determine）
turnover *n.* 营业额
eliminate *vt.* 消除，排除（get rid of）

valuation *n.* 估价（evaluation）
conclusive *adj.* 确定的，决定性的（decisive）
exclusive of 不包括（excluding）
VAT 增值税

疑 难 词 句

1. It says nothing about quality of materials or workmanship, nor does it indicate satisfaction with the work done to date.

译文：它既不涉及工程的材料质量或工艺水平，也不表示对至今所完成的工程满意。

9.3 Variations

In order to change the specification of the work, a contract would, in principle, have to be re-negotiated. To avoid this, most contracts include a clause which enables the employer's design team to vary the specification. Such provisions are usually called variations clauses.

9.3.1 The need for variations

There are three ways in which a variation might occur:

1. Clients may change their minds about what they asked for before the work is complete.

2. Designers may not have finished all of the design and specifications before the contract was signed.

3. Changes in legislation and other external factors may force changes upon the project team.

Although these three origins are very different, construction contracts tend to ignore these differences and deal with all variations in the same way. Clearly, buildings are so complex as to require changes to be made before they are completed. Client-instructed variations are the prerogative of the client. The second group of variations arises because it is rare for design to be completely detailed at the time of tender; changes may have to be made simply to make the building work. However, there are cases where variations have to be instructed simply because designers have failed to complete their task and in their haste to get a project out to tender they leave large portions of the work unspecified. The presence of a variation clause enables dilatory design teams to delay finalization of the design until late in the project. Such behaviour jeopardizes the success of general contracting because of the assumptions on which it is based. It is bad practice and should be avoided as far as possible. The third source of variations, external factors, is typically beyond the control of any of the parties.

If there were no provision in the contract for varying the work to be done, then any attempt by the employer to vary it would require the agreement of the contractor. The lack of a provision for variations would effectively enable the contractor to negotiate a new price for the whole contract every time the employer tried to make a change. This apparent difficulty is not as rigorous as it sounds. The absence of a variations clause undoubtedly makes it difficult to vary the terms of the contract but it is at least possible that the courts would imply a term allowing minor variations to be made. In any event, it would of course be most unusual for a contractor to attempt to refuse to carry out small changes and even less likely that the contractor would go to court over an attempt to impose them.

By inserting a clause which allows for changes to be made to the works as they are being built, the employer, through the contract administrator, can alter the works as and when necessary. The purpose of the variation clauses is to allow such changes to be made, and also to permit any consequential changes to be made to the contract sum.

It is always possible for a contract to include a clause that fixes express limits on the amount of variations. In any event, it must be borne in mind that the existence of a variations clause does not entitle the employer to make large scale and significant changes to the nature of the works, as these are defined in the tender document. In particular, variations which *go to the root of the contract* are not permissible. If the tender document states that 8 dwelling houses are to be built, then a variation altering this to 12 would possibly be construed as going to the root of the contract. However, if the tender document states that the contract is for 1008 houses, then a variation changing this to 1012 would not go to the root of the contract, because it would be a minor change in quantity. What is probably more important is that if the contract is for the erection of a swimming pool, a variation which attempts to change it to a house would clearly be beyond the scope of the contract.

9.3.2 Definition of variations

The definition of variation is a wide one, applying to both the content of the work and the method of doing it. Indeed, at first glance the definition appears almost unlimited in scope, since it makes repeated reference to 'any' work. However, bearing in mind what was said earlier about the tender document, and not changing the nature of the contract, this is clearly not the true position.

Alteration of the works

The basic definition includes any alteration or modification of the design, quality or quantity of the works as shown on the contract drawings and described by or referred to in the contract bills. Such alteration or modification includes additions, omissions or substitutions, which may possibly refer to the situation where a thing is varied after it has been completely built.

This part of the definition of a variation specifically includes the alteration of the kind or standard of any of the materials or goods to be used in the works. Such changes necessarily involve the contractor's head office in expenditure, such as the additional time needed to cancel orders, find alternative suppliers and re-order materials and goods. Detailed records may be needed to prove this type of expenditure, or alternatively it may be calculated on a pro-rata basis.

The third aspect of this part of the definition is the removal from the site of work or materials which are there for the proper purposes of the contract. This removal, however, in some other contracts by contrast, also refers to the removal of things which are not in accordance with the conditions.

Alteration of working methods

In addition to allowing the works themselves to be altered, most contracts allow for variations in the means of achieving those works. The means available to the contractor may be seriously affected by the site conditions, working space, hours of work or the sequence in which work is to be carried out. Unlike the permitted variations to the works, however, these definite and specific matters of working method are fairly limited in the way in which they can be varied. It must also be borne in mind that exercising further restrictions in these things could result in substantial money claims by the contractor.

In accordance with the common law position, the contract provides specifically that none of the contractor's work can be taken away and given to others while the contract is current.

9.3.3 Issuing variations

In order for such an instruction of variation to be valid under the contract, it must be issued in accordance with the contract. This means, for example, that it must be in writing. It also means that the contractor has the right to object on reasonable grounds to a variation.

As well as issuing an instruction which may require a variation, the contract administrator may sanction a variation that has been made by the contractor. The contractor normally has no authority whatsoever to vary anything (except for the execution of emergency work as required by the contract). For example, the contractor cannot substitute higher quality work or materials than those specified. A contractor who does vary anything then can be instructed to remove it. This power for the contract administrator to sanction an unauthorized variation by the contractor is thus an important one, since it allows the contract administrator easily to pick up and deal with minor items which are varied by the contractor and not confirmed by the contract administrator.

9.3.4 Valuing variations

All variations to the contract are to be valued by the quantity surveyor. Many contracts provide that variations and work carried out by the contractor in expenditure of provisional sums and work for which an approximate quantity is included in the contract bills, shall be valued in accordance with the relevant provisions. The rules to be observed in valuing variations may be summarised as follows:

- Work which can be properly valued by measurement:

(a) Work similar to that set out in the contract bills, executed under similar conditions and with no significant change in the total quantity, shall be valued at bill rates. This includes work where the approximate quantity in the contract bills is a reasonably accurate forecast of the quantity of work required.

(b) Work similar to that set out in the contract bills but not executed under similar conditions or where there is a significant change in quantity, shall be valued on the basis of those in the contract bills but with a fair allowance being made for the differences in conditions and/or quantity. This includes work where the approximate quantity in the contract bills is not a reasonably accurate forecast of the quantity of work required.

(c) Work not similar to that set out in the contract bills shall be valued at fair rates and prices.

- Omissions shall be valued at the rates and prices contained in the bills.
- When valuing the foregoing, the following must be taken into account:

(a) Measurement must be in accordance with the method of measurement used for the preparation of the bills of quantifies.

(b) Allowance must be made for any percentage or lump sum adjustment in the contract bills.

(c) If appropriate, allowance must be made for any additional or reduced preliminaries.

- Work which cannot properly be valued by measurement shall be valued on a daywork basis.
- If any variation substantially changes the conditions under which other work is executed, then that other work shall be revalued as if it was itself a variation.
- If the variation does not involve additional or substituted work or straightforward omissions, or if the valuation cannot reasonably be effected by the application of these valuation rules, then a fair valuation must be made.

When valuing variations, it is important to note that, where a variation causes, in the case of other work, either a change in the working conditions, or a significant change in the quantities or in the conditions under which this other work is carried out, then such changes must also be taken into account in the valuation of the variation.

It is important to note that allowance must be made, in valuing variations, for percentage or lump sum adjustments which have been made in the contract bills, for example for profit additions, and, where appropriate, for adjustment also to be made in respect of the additional or reduced preliminaries requirements.

It is not necessary to take account, when valuing variations, any effect which the variations may have on the regular progress of the work, or any direct loss and/or expense which the contractor may have incurred as a result of the variation and which he is unable to recover through the valuation or any other provision of the contract.

The valuation of variations to nominated sub-contract works, including the valuation of work carried out against provisional sums included in the sub-contract, is to be made in accordance with the relevant provisions of the sub-contract. The valuation rules in the sub-contract conditions are similar in principle to those of the main contract referred to above. Any variations in the sub-contract works are then valued in accordance with the terms of the sub-contract, not in accordance with the terms of the main contract.

It is interesting to note that, when the contract rules for valuing variations are exercised, the quantity surveyor has a unilateral responsibility - the contractor is not involved, apart from being entitled to be present when any measurements are made. Should the contractor not be satisfied with the quantity surveyor's valuation, his only formal recourse is to use the dispute resolution procedures set out in the contract. Practically, of course, the quantity surveyor usually works closely with the contractor's surveyor, so that a dispute does not arise and, hopefully, an agreed final account is eventually produced.

9.4 Fluctuations

We may now look briefly at fluctuations, which are the third method by which the contract sum may validly be adjusted. The purpose of a fluctuations clause is to provide a mechanism for reimbursing contractors for changes in input prices over which they have no control at all. Even if a contractor has caused the project to take longer than planned, the changes in market prices for supplies will be considered by a valid fluctuations clause, unless the clause expressly provides otherwise. Many contracts specifically provide that a contractor who is in delay shall not benefit from the fluctuations clause.

Most contracts contain the fluctuations clauses. Usually there are three fluctuation schemes as follows:

- *Levy and tax fluctuations*. This clause applies to items which are affected by the government and are thus completely beyond both the control and the prediction of the contractor. The elements to which the clause applies fluctuations are labour, materials and goods, electricity and fuels, but only to the extent that they are affected by tax etc. In some cases they are only covered as far as they have been listed in the contract documents. There are no methods for calculation set out in the clause, but obviously it would be necessary to take into account man-hours, quantities of materials and other directly ascertained costs. It is not intended that this scheme should change the amount of contractor's profit.

- *Labour and materials cost and tax fluctuations.* This clause includes all the government-related items that are covered by the contract; it adds in labour and materials fluctuations. This covers the market costs of input, such as wage rates and prices of materials.

- *Use of price adjustment formula.* This is a completely different type of calculation. It incorporates by reference a set of formula rules which define a technical financial calculation based on a wide variety of categories. The whole of the works is divided up into financial categories and a monthly published bulletin gives indices by which each sum should be multiplied. In order for this to work, it is necessary for the contract bills to reflect the categories used by the fluctuation categories. Since the index numbers reflect the market situation each month, they are deemed to include the matters covered by the contract. The purpose of this method is to reduce the amount of calculation which has to be undertaken by the project team.

These different types of fluctuations are referred to as limited fluctuations and full fluctuations respectively. In addition, it is of course possible to have no fluctuations provisions at all especially for a small project under stable conditions.

9.5 Retention money

It is common practice in the construction industry to withhold a small proportion of payments to a contractor until the work has been completed satisfactorily. These amounts are referred to as retention and are usually deducted from each interim certificate.

The retention fund is intended to be available to the employer for the purpose of rectifying, or inducing the contractor to rectify, any defects in the work appearing during the defects liability period. This period lasts from the date of practical completion as certified by the contract administrator for whatever length of time is specified in the appendix to the contract. One year is common.

Many contracts contain a specific provision to illustrate the operation of a retention scheme. The effect of those provisions is that, on the issue of every interim certificate, the employer is entitled to deduct the agreed retention percentage. This is deducted from the value of work which has not yet reached practical completion and from the value of any materials included in the certificate. The employer is also entitled to make equivalent retention from sums due to nominated sub-contractors.

It is further stated that the retention percentage referred to will be whatever figure the parties have entered in the appendix to the contract. The common figure is 5%, although some contracts suggest that, if the contract sum is expected to exceed a certain amount, it should not be more than 3%. Further, it is common to provide a ceiling beyond which no more will be retained, although many contracts contain no such provision.

After the contract administrator certifies that practical completion has been achieved, but before the issue of a Certificate of Completion of Making Good Defects, One-half of the accumulated retention money currently held by the employer will be released to the contractor. The remainder will be released by the interim certificate issued either at the end of the Defects Liability Period or upon the issue of the Certificate of Completion of Making Good Defects, whichever is the later.

Words, Phrases and Expressions

origin *n.* 来源，原因（source）
prerogative *adj.* 特权（privilege）
dilatory *adj.* 延误的（delayed）
jeopardize *vt.* 危及，损害（harm）
vary *vt.* 使变化（change）
rigorous *adj.* 严厉的，严格的（severe）
consequential *adj.* 作为结果的，由此产生的（resultant）
construe *vt.* 理解（notes，understand）
erection *n.* 建造（construction）
modification *n.* 修改（revision）
alternative *n.* 另外的（another）
pro-rata *adj.* 按比例的
definite *adj.* 明确的，确定的（clear，certain）
object *vi.* 反对（oppose）
sanction *vt.* 批准，认可（approve）
approximate *adj.* 近似的，大约的
allowance *n.* 补贴，补助（subsidy）
daywork *n.* 计日工
fair valuation 合理的估价
unilateral *adj.* 单方的
recourse *n.* 求助对象，采用的办法（resort）
fluctuations clause 价格波动条款（escalation clause）

bulletin *n.* 公告

indices *n.* 指数（复数）

equivalent *n.* 相等的，相当的（equal）

ceiling *n.* 上限（upper limit）

疑 难 词 句

1. However, there are cases where variations have to be instructed simply because designers have failed to complete their task and in their haste to get a project out to tender they leave large portions of the work unspecified.

译文：不过，有些时候仅仅是由于设计师没能完成他们的任务，而在大部分工作都没确定的情况下为了投标仓促地拿出一个项目，从而使得后来不得不要变更。

2. Such behaviour jeopardizes the success of general contracting because of the assumptions on which it is based.

译文：这种行为会危及总承包方式的成功，因为总承包是基于该假定设计的。

3. The absence of a variations clause undoubtedly makes it difficult to vary the terms of the contract but it is at least possible that the courts would imply a term allowing minor variations to be made.

译文：没有变更条款无疑会使合同条件的变更变得困难，但法庭至少有可能把允许少量的变更作为默认条款。

4. Such alteration or modification includes additions, omissions or substitutions, which may possibly refer to the situation where a thing is varied after it has been completely built.

译文：这种变更或修改包括增加、删减或替代，并且可能涉及到某项工程完工后又需要变更的情况。

5. The means available to the contractor may be seriously affected by the site conditions, working space, hours of work or the sequence in which work is to be carried out.

译文：承包商能采用的施工方法可能会受到现场条件、工作空间、工作时间或工作顺序的严重影响。

6. Many contracts provide that variations and work carried out by the contractor in expenditure of provisional sums and work for which an approximate quantity is included in the contract bills, shall be valued in accordance with the relevant provisions.

译文：很多合同都规定：用暂列金支付的变更和承包商完成的工程以及工程量清单中有近似工程量的工程都应该按照有关条款的规定估价。

7. If any variation substantially changes the conditions under which other work is executed, then that other work shall be revalued as if it was itself a variation.

译文：如果任何变更使得其他工程项目的实施条件遭受到了重大改变，那么这些其他工程项目就应该作为变更而重新估价。

Chapter 10 Role of the Contract Administrator

The purpose of employing an architect, engineer, or other professional person on a building project is to give the employer the benefit of that professional's skill and experience. Traditionally, the person appointed has taken responsibility for two separate functions: translating the employer's needs into drawings, specifications and the like through the processes of briefing and design, and then supervising the work of actual construction. This is done to ensure that the work complies with the designer's intentions and satisfies standards of workmanship and quality.

Although the functions of design and contract administration are still frequently performed by the same person, this is by no means essential. Indeed, under some forms of building procurement such as construction management, it is unlikely to be the case. In any event, however, the functions of a contract administrator (whether described as architect, engineer, supervising officer, project manager or whatever) are completely separate from those of a designer.

As we shall see, a contract administrator fulfils two rather different roles. First, there are those duties (such as providing necessary information to the contractor) which are carried out as agent of the employer. Second, there are certain decision-making functions (such as the certification of work properly carried out) in which the contract administrator is required to act fairly between the parties and exercise independent judgement.

10.1 Contract administrator as the employer's agent

A contract administrator who is employed to supervise the carrying out of building works may in certain respects be regarded as an agent of the employer. A number of important issues result from this, concerning especially the extent to which the contract administrator can bind the employer by his or her actions, and the scope of the authority exercised by the contract administrator on behalf of the employer.

10.1.1 Extent of powers

In acting for the employer, there are limits to what a contract administrator can do in terms of forming or varying contracts, instructing work to be suspended and in delegating authority to others.

Contracts and variations

An agent can of course be expressly authorized to do anything on behalf of a client.

However, the important question of law concerns the extent of authority which will be implied where nothing specific is said. This is important because, where a contract administrator acts without any authority, the employer will not be bound by what is done. In such a case the third party concerned may sue the contract administrator personally for damages for breach of warranty of authority.

As a general rule, the courts take a rather restricted view on this question. In particular, it has been consistently held that, except in very unusual circumstances, a contract administrator has no authority to create a direct contract between the employer and a sub-contractor. Thus, where an architect promised a sub-contractor that the employer would pay directly for the work, it was held that the employer was under no obligation to do so.

A similarly cautious approach is seen in cases concerning the alteration of an existing contract. Unless the contract administrator acts within the terms of a variations clause, there is no power to change what was originally agreed.

As mentioned above, a contract administrator who exceeds his or her authority risks being held personally liable to a third party with whom he or she deals. In addition, the law of agency contains another trap for the unwary. This is that any agent who signs a written contract on behalf of a client will be treated as a party to it and thus personally liable, unless the contract itself makes it clear that it is signed merely 'as agent'.

Suspension of work

As a general principle, neither the employer nor the contract administrator has any legal right to order the contractor to suspend work. Once the contract work has commenced, it is the contractor's right and duty to carry it through in a regular fashion, and the employer's duty to do nothing which will hinder the contractor in this. As a result, an unjustified order to suspend work, given by the contract administrator, will amount to a breach of contract for which the contractor may claim damages.

However, this general principle is substantially modified by most of the major standard form contracts. Many contracts, for example, give the engineer an unfettered discretion to order the suspension of the whole or any part of the works. Depending on the reason for this order, the contractor may or may not be entitled to claim any costs incurred as a result of it. Once such a suspension has lasted for three months, the contractor is entitled to request permission to resume. If this permission is not given within 28 days, then the contractor may treat the suspended part as omitted from the contract or, if the suspension order relates to the whole, may treat the contract as terminated.

It is perhaps worth mentioning that most standard form contracts permit the employer or contract administrator to suspend all or part of the work. However, although the contractor is entitled to an extension of time and to financial compensation where this occurs, there is no provision for the contractor to terminate in the case of a long suspension. In such circumstances, a contractor wishing to bring the contract to an end must show that the employer's conduct amounts to a breach of contract which entitles the contractor to terminate the contract.

As far as sub-contracts are concerned, these will almost invariably give the main contractor a power of suspension which are similar to the power of the employer or architect under the main contract. It is indeed essential for the protection of the contractor that this should be so for, if it is not, the contractor may incur liabilities to the sub-contractor which cannot be passed on. It is worthwhile to note that this main contractor's power of suspension of the sub-contractor's work should be included in the sub-contract.

Delegation of authority

As a general principle, an agent is expected to act personally for a client, and not to delegate the task to a sub-agent. However, authority to delegate may be given expressly or by implication. In particular it seems that, where a contract administrator is instructed to invite tenders for the contract work on the basis of bills of quantities, authority to appoint a quantity surveyor will be implied. What is not clear, however, is whether this creates a direct contract between the employer and the quantity surveyor, or whether it simply means that the contract administrator can recover from the employer sums paid to the quantity surveyor. Similar authority to appoint may be implied wherever the building contract provides that variations are to be measured by a quantity surveyor. In such circumstances, it is clear that different parts of the 'contract administration' are to be carried out by different people.

Besides these implications, some contracts made express provision for the appointment of a quantity surveyor by the employer after consultation with the architect. Some other contracts give the power to appoint quantity surveyors to either the employer or the architect, but only with the a separate agreement.

At this point mention should also be made of the possibility that an employer will wish to appoint a clerk of works, who will be permanently on site to act as the 'eyes and ears' of the contract administrator. Since such a person is almost invariably appointed by the employer directly, the division of supervisory duties between contract administrator and clerk of works can cause problems.

The general principle is that, while matters of detail may be left to a properly briefed clerk of works, the contract administrator remains responsible for seeing that important matters of design are properly carried out. Thus, where a clerk of works, for corrupt purposes, allowed the contractor to deviate from the design in laying concrete in a way which led to dry rot in the ground floor, the architect was held liable to the employer. It was said that the architect might justifiably have supervised the laying of concrete in the first building and then left the remainder to the clerk of works; here, however, the architect had not supervised or inspected at all.

Notwithstanding this decision, it is clear that failure of proper supervision by a clerk of works is something which may affect the employer. Thus, for example, where a clerk of works negligently failed to notice defects in the way that artificial stone mullions were fixed, the damages which the employer was awarded against the architects for negligence were reduced by 20%, on the ground of contributory negligence by the clerk of works for which the employer was responsible.

10.1.2 Functions and duties

In carrying out 'administrative' functions under a building contract, the contract administrator naturally owes a duty of reasonable care and skill to the employer. If that duty is breached, the contract administrator will be liable in damages to the employer for any resulting loss. For example, there was a case that a contractor was in serious breach of the obligation to maintain regular and diligent progress. The Court of Appeal held that the architect should have served a notice determining the contractor's employment, and was liable to the employer for various losses which flowed from the fact that no such notice was served.

In considering the contract administrator's functions, two areas of special importance relate to advice and to supervision of the work.

Advice to the employer

The kind of matters on which it is reasonable to expect a contract administrator to give advice will naturally vary with the type of project, the professional background of the contract administrator and so on. However, three cases in which architects were held to have been negligent may give some indication. In the first of these, the architect greatly underestimated the extent of work required to reinstate an old property. As a result the client, having purchased the property, could not afford to refurbish it. In the second case, the architect, in preparing estimates of the cost of a project, completely overlooked the effects of inflation. Once again the client was forced to abandon the project. In both these cases it was held that the architect was not entitled to any fees, since their work was worthless.

In addition to matters of cost, a contract administrator may well advise the employer on the appointment of particular contractors or sub-contractors. This is especially likely where a tendering process has taken place. In such circumstances, once again, a duty of care is owed. Thus, where an architect recommended a particular contractor as 'very reliable', the architect was held liable for losses suffered by the employer when the contractor executed defective work and then became insolvent.

Another area of advice for which employers depend upon the contract administrator is advice about the rights and duties within the building contract. For example, once in a project dispute the architect was held negligent because the employer under a minor works contract had not been advised by the architect to take out insurance for the works. As a result, damage to the works caused losses far in excess of the contract value. It is extremely important that a contract administrator explains the terms of the building contract to the employer before the contract is executed.

Instructions to the contractor

There is no general right under building contracts for contract administrators to issue instructions to contractors. It is only where there is an express condition that this can happen. However, in practice standard form contracts always contain such a clause. This is because the duration and complexity of construction projects are such that conditions are likely to change, and it is recognized that it may thus not be possible to deal in advance with every eventuality which may arise.

Contractual clauses dealing with this matter are often very widely drafted. Some contracts, for example, list a large number of matters on which the project manager may issue instructions. FIDIC Conditions is hardly more limited; having restated the contractor's basic obligation to build in accordance with the contract, it requires the contractor to comply with and adhere strictly to the Engineer's instructions and directions on any matter connected therewith.

The position under some other contracts is more restricted that the contractor's duty of compliance only applies to written instructions which the contract administrator is expressly empowered by the conditions to issue. If there is any doubt about this, the contractor can ask the contract administrator to specify in writing the provision under which the instruction is issued, following which any further dispute can be taken immediately to arbitration. Assuming that an instruction is justified, the contractor must comply with it within seven days, failing which the employer is entitled to have it carried out by someone else at the contractor's expense. On the other hand, where compliance with an instruction involves the contractor in delay, it is possible for an extension of time to be claimed. As to any extra cost involved, the contractor will normally be reimbursed for this either through the provisions for valuing variations or through those relating to loss and/or expense.

Information to the contractor

Apart from issuing instructions, the contract administrator has an important function as a source of relevant information to the contractor. Many contracts require the contract administrator to furnish the contractor with one copy of the contract documents certified on behalf of the employer, two further copies of the contract drawings and (where relevant) two copies of the unpriced bills of quantities. These documents are to be furnished immediately after the execution of the contract. And further the contract administrator shall provide two copies of any further information necessary to enable the contractor to complete the works in accordance with the conditions.

Having thus provided the contractor with the basic information required, the contract administrator must keep this information up to date. Many contracts provide that 'as and when from time to time may be necessary' the contract administrator shall provide such further drawings or details as are reasonably necessary either to explain the contract drawings or generally to enable the contractor to carry out the work.

It is worthwhile to note that one important aspect of the provision of information is as to the determination of any levels which may be required for the execution of the works. Some contracts make this the responsibility of the contract administrator, who has further to furnish the contractor with sufficient accurately dimensioned drawings to enable the works to be set out at ground level. Provided this is duly done, the responsibility for any errors in setting out rests on the contractor. This will normally involve amending the errors, but the contract empowers the employer and contract administrator to deal with the matter instead by means of an 'appropriate' deduction from the contract sum.

An important question in this context is the extent to which a contract administrator may incur personal liability to a contractor for inaccuracies in information provided. It has long been settled that in drawing up plans, an architect or engineer does not guarantee they are practicable, nor that bills of quantities are accurate. In principle, therefore, contractors who wish to tender on the basis of such information must satisfy themselves as to its soundness. However, there are limits to this principle. For example, an architect who fraudulently gives inaccurate information at the pre-tender stage will be liable to the contractor. This is so even if the contract provides that the contractor must not rely on any representation contained in the plans.

Inspection and supervision

Where a qualified architect is appointed to be contract administrator, it is likely that the terms of engagement make it clear that while the architect will visit the site at intervals, there is no requirement to make 'frequent or constant inspections'. If such inspections are necessary, a clerk of works or resident architect is to be appointed. Indeed, it appears to exclude altogether the architect's liability for failure to supervise. This is because it provides that the employer will hold the contractor or consultant and not the architect responsible for the proper execution of their respective work. However, it appears that such wording will not override the architect's traditional duty to check general compliance.

Some contracts clearly recognize that the contract administrator will inspect all work executed. The contractor's basic obligation is to produce the building according to the contract documents. While the contract administrator will naturally wish to check that this is indeed being done, the actual need to inspect relates only to those matters which are required to be to the contract administrator's satisfaction. In this context it may be remembered that any such stipulation means that work must be to the *reasonable* satisfaction and not the *absolute* satisfaction of the contract administrator.

As part of the function of inspecting the contractor's work, the contract administrator is given considerable powers to carry out tests of any materials and to order that work which has been covered up by other work should be opened up for inspection. The cost of making good and restoring the covering work, and any work found to be defective, depends on the status of the inspected work. If no defects are discovered, then the contractor must be paid for the cost of opening up and making good as if this was due to a variation.

If the inspected work is defective, then this is to be made good at the contractor's expense. What is more, once defects have been discovered, most contracts entitle the contract administrator to demand further tests. Provided that any such demands are in accordance with the contract, the cost of them is borne by the contractor whether or not any further defects are discovered.

The contract administrator clearly owes a duty of care and skill to the employer to detect bad workmanship and defects. However, due to the limited nature of the duty to inspect it cannot be said that every failure of detection will amount to negligence. Further, while it is sometimes said that a similar obligation is owed to contractors, this seems unlikely; they are responsible for their own monitoring of quality. The most that can be said is that, where the state of the work is positively dangerous, there may be an obligation to warn the contractor of this. Of course, common sense suggests that, regardless of the strict legal position, a contract administrator who is aware of some defective work should certainly communicate this to the contractor.

Mention of danger leads to the more general question of who is responsible for site safety. This in turn raises the issue of whether the contract administrator has power to control the contractor's method of working. A power of this nature is undoubtedly given to the engineer.

10.2 Contract administrator as independent adjudicator

As we mentioned before, a contract administrator has a significant part to play in exercising judgment and reaching decisions on various matters under the contract. In so doing the contract administrator acts, not as the agent of the employer, but as an independent professional.

10.2.1 Certification

The most important aspect of these decision making powers relates to the issue from time to time of what are called certificates. These have been defined as 'the expression in a definite form of the exercise of the judgment, opinion or skill of the engineer, architect or surveyor in relation to some matter provided for by the terms of the contract'. However, this does not mean that every expression of opinion or decision given by the contract administrator will amount to a certificate. It will only be a certificate if it is so described in the contract, or can be so treated by implication.

The general law of building contracts does not require a certificate to be given in any particular form. Indeed, it may be given orally unless the contract provides otherwise. However, most contracts, though not necessarily expressed, imply that the certificates must be issued in writing.

Types of certificate

There are three main types of certificate found in construction contracts:

1. *Interim certificates*. These are issued at intervals as the work proceeds, and their issue entitles the contractor to be paid a certain proportion of the contract price. The period between interim certificates is whatever is stated in the appendix. If none is stated then it is one month, which is the usual period. The importance of the interim certificate lies in that the regular flow of cash can be critical to a contractor's survival.

The amount to be included in an interim certificate should cover the value of work done and materials delivered to date, plus the value of certain off-site materials. The contract places the responsibility for carrying out interim valuations upon the quantity surveyor. However, it is important to note that the QS should not do more than merely carry out the valuations; it is for the contract administrator to calculate what is due and to issue the interim certificates.

2. *Final certificates*. The final certificate can signify the contract administrator's satisfaction with the work, or the amount which is finally due to the contractor, or both of these things. The extent of its effect depends on the terms of the contract. Under many contracts, the obligation is in general to issue the final certificate within a time limit of the end of the defects liability period. In practice, it usually takes much longer than this specified period to issue final certificates. Indeed, on some projects final certificates are only issued some years after completion, rather than months. There are two potential reasons for this situation. First, the last portion of the retention may be too small to worry a contractor, especially if the employer has deducted some part for any reason. Second, consultants may find that other projects, at earlier stages in their development, are more lucrative than those which can yield only the final part of the fee account.

3. *Certificates recording* an event. In addition to confirming that a sum of money is due to the contractor, certificates may be needed to confirm that a certain event has occurred (or has not occurred). The use of this kind of certificate varies from one form of contract to another, but a good example of their use and importance may be summarized as below, and many contracts require the contract administrator to issue the following certificates:

- *Certificate of non-completion*. This records the contractor's failure to complete the works by the completion date. Its importance is that it triggers the contractor's liability to pay liquidated damages.
- *Certificate of practical completion*. This records the contract administrator's opinion that practical completion of the works has been achieved. The contractor's liability for liquidated damages ceases, one-half of the retention money is released and the Defects Liability Period begins.
- *Certificate of completion of making good defects*. This records the contract administrator's opinion that defects appearing within the defects liability period and notified to the contractor have been duly made good. The contractor is then entitled to the

remainder of the retention money.

10.2.2 Other decision-making functions

Although the issue of certificates is the most important aspect of the contract administrator's 'independent' role, it is not the only one. Construction contracts may also use other forms of words, such as requiring the contract administrator to 'make decisions' or to 'give opinions'.

An example of 'certification by another name' is provided by many standard form contracts, which state that the employer may take possession of part of the works before practical completion is achieved. Where this occurs, the contract administrator is required to issue a 'written statement' identifying the part taken into possession. This statement is then treated for most purposes as if it were a certificate of practical completion for the relevant part.

An extremely important function of this kind is found in the relevant clause of many contracts. This provides that any dispute whatsoever between the employer and the contractor arising out of the contract or the work must be referred in writing to the engineer for settlement. The engineer must then give a decision in writing on the dispute. Only when this has been done (or when the engineer has failed to give a decision within a prescribed period) can the matter be taken to arbitration.

10.2.3 Liability for negligent decision-making

A question which has been the subject of important litigation in recent years is whether a contract administrator can be held liable to either of the contracting parties for any decisions.

Liability to the employer

It was for many years believed that, in issuing certificates, a contract administrator acts in a 'quasi-judicial' capacity. By this is meant that, in exercising judgement about the quantity of work done, and making decisions about how much the contractor should be paid, the architect is acting in a manner similar to that of an arbitrator. This led to the conclusion that the contract administrator should enjoy the same immunity from claims in negligence as an arbitrator (which is itself based on the immunity of a judge).

In particular, it suggests that the contract administrator must notify the quantity surveyor in advance of any work which is regarded as not properly executed, so that it can be excluded from the quantity surveyor's valuation. Then this portion will not be included in the interim certificate.

It appears that the duty of care which a contract administrator owes to the employer applies, not only to certification as such, but also to the other decision-making functions. Liability may thus arise where the contract administrator, by negligently granting extensions of time to which a contractor is not in truth entitled, causes the employer to lose liquidated damages.

Liability to the contractor

Once it was held that a contract administrator could be liable to the employer for negligent over-certification, the question which naturally arose was whether there would be an equivalent liability to the contractor for negligent under-certification. This, it was said, would cause serious losses because, even if the contractor ultimately recovered the correct amount from the employer, there would still be an expensive interruption of the contractor's cash flow.

Words, Phrases and Expressions

suspend *vt.* 暂停（pause）
unwary *adj.* 不警惕的，易受骗的
fashion *n.* 方法（way, method）
unfettered *adj.* 无拘无束的（unconstrained）
resume *vi.* 重新开始，恢复（restart）
incur *vt.* 遭受，招致（suffer from）
clerk of works 现场监工，工程管理员
division *n.* 分开，界限，分裂（separation）
corrupt *adj.* 腐败的
concrete *n.* 混凝土
damages *n.* 损害赔偿金（compensation）
contributory negligence 互有过失，原告的疏忽
flow from 起因于，产生于
underestimate *vt.* 估计不足，低估
refurbish *vt.* 重建，改建（rebuild）
take out *vt.* 获取，办理（obtain）
furnish *vt.* 提供（provide）
dimension *n.* 尺寸（size）
amend *vt.* 修订，改正（correct）
practicable *adj.* 可行的，适用的（workable）
soundness *n.* 完善，正确（correctness）
at intervals 不时，相隔一段距离（occasionally）
considerable *adj.* 相当大（多）的（substantial）
status *n.* 状况，状态（state）
demand *vt.* 要求（require）
detect *vt.* 发现，发觉（discover）
positively *adv.* 肯定地，绝对地（certainly）

regular *adj*. 连续的（continuous）
survival *n*. 生存
signify *vt*. 表示…的意思（mean）
lucrative *adj*. 获利多的，赚钱的（profitable）
trigger *vt*. 引发（start）
quasi-judicial *adj*. 准司法性的
immunity *n*. 豁免，免除（being exempt）
grant *vt*. 给予（administer）
over-certification *n*. 过度签证（超过实际工程量的签证）
under-certification *n*. 不足签证（低于实际工程量的签证）
interruption *n*. 中断（halt）

疑 难 词 句

1. A number of important issues result from this, concerning especially the extent to which the contract administrator can bind the employer by his or her actions, and the scope of the authority exercised by the contract administrator on behalf of the employer.

译文：由此引发了一些重要的问题，尤其是合同管理者在多大程度上能通过他们的行为约束业主，以及他们代表业主行使权力的范围等。

2. However, the important question of law concerns the extent of authority which will be implied where nothing specific is said.

译文：但是，当没有明确说明的情况下，默认的授权程度（范围）是一个重要的法律问题。

3. This is that any agent who signs a written contract on behalf of a client will be treated as a party to it and thus personally liable, unless the contract itself makes it clear that it is signed merely'as agent'.

译文：即：任何代表业主签订书面协议的代理将被当成该协议的当事方，并亲自承担责任，除非该合同本身明确表明其签署仅仅是以代理的名义。

4. However, although the contractor is entitled to an extension of time and to financial compensation where this occurs, there is no provision for the contractor to terminate in the case of a long suspension.

译文：然而，虽然承包商有权因此延长工期并获得经济赔偿，但并没有规定承包商在工程长期暂停后有权终止合同。

5. The general principle is that, while matters of detail may be left to a properly briefed clerk of works, the contract administrator remains responsible for seeing that important matters of design are properly carried out.

译文：一般原则是：具体的小事可以交给比较了解现场情况的工程管理员来做，合同管理者则负责确保设计方面的重要事宜得到适当的贯彻。

Notes：'see that' 指的是"监督或确保某项工作得到顺利的执行"。

6. The kind of matters on which it is reasonable to expect a contract administrator to give advice will naturally vary with the type of project, the professional background of the contract administrator and so on.

译文：合理地期待合同管理者提出咨询意见的这类事宜将很自然地随着项目的种类、合同管理者的专业背景等因素而发生变化。

7. As to any extra cost involved, the contractor will normally be reimbursed for this either through the provisions for valuing variations or through those relating to loss and/or expense.

译文：关于所涉及的额外费用，承包商通常会根据变更估价条款或有关损失或费用的条款得到偿还。

8. Liability may thus arise where the contract administrator, by negligently granting extensions of time to which a contractor is not in truth entitled, causes the employer to lose liquidated damages.

译文：当合同管理者因疏忽而给予承包商事实上本无权获得的延长的工期，使业主因此丧失了误期损害赔偿金时，合同管理者可能会因此承担责任。

Chapter 11 Subcontracting and Nomination

One prominent characteristic of construction industry is the practice of subcontracting, which provides opportunity and benefit to both contractors and subcontractors and high efficiency to the industry as a whole. Actually, subcontracting runs far beyond construction industry, it is also widely used in information technology and information sectors of business and so on. In this chapter, we will focus on how subcontracting works in construction business.

11.1 Outline of Subcontracting

11.1.1 What is construction subcontracting

Subcontracting is a type of work contract that seeks to outsource certain types of work otherwise would be undertaken by general (main) contractors (sometimes short as MC) to other independent companies (subcontractors). A subcontractor is an individual or in many cases a business that signs a contract to perform part or all of the obligations of another's contract (main contract or general contract), that is to say, the subcontractors are supposed to perform identifiable services within the scope of the main contract. Subcontracting is probably the most prevalent in the construction industry, where general contractor often subcontract a great portion of work, for example, supplying materials, plumbing, cooling and heating, drainage, roofing, electrical work, drywall, painting, the services of a carpenter, bricklayer or carpet layer, plastering, flooring, and structural steel erection and so on.

11.1.2 Reasons for subcontracting

Subcontracting offers a number of advantages:

a. Subcontracting sometimes is applied when the general contractor does not have the time or skills to perform certain tasks. In some cases, a general contractor may virtually be used as the construction manager or supervisor. In that case, subcontracting accounts for all of the physical work done and the only responsibility of general contractor is to approve the contracts, keep the project within budget and inspect the work.

b. Subcontracting is cost saving. First of all, subcontracting, more often than not, allows work on more than one phase of the project to be done at once, which instead of time consuming, often leads to a quicker completion. Secondly, with subcontractors being particularly skilled and experienced in certain fields, it is often much cheaper and quicker for them to complete their part of job than a general contractor does. Finally, in the case that a subcontractor have been working with a general contractor for many years and on more than one project, the already formed long term relationship may create a saving for both parties.

c. By subcontracting, general contractor may receive the same or even better service than otherwise could be provided by himself, and at lower overall risk. In fact, many general contractors exploit subcontracting mainly to shift liability risks. In this way, subcontractors are often contractually required (e. g. by fixed price contracts) to absorb much of the uncertainty and risk faced by the general contractor.

11. 1. 3 Types of subcontractors

When we talk about subcontractors, we mostly refer to domestic subcontractors. But, there are other types of subcontractors in practice, with some chosen by general contractors and some nominated by employer or even government.

a. **Domestic subcontractor**

Domestic subcontractor (sometimes short as DSC) is decided and employed by the general contractor, with employer playing no part other than simply giving consent. So the appointment of the subcontractor is treated as a "domestic issue" that is entirely up to MC. A subcontractor contracts with the general contractor to supply or fix any materials or goods or execute work which is part of the main contract and is fully responsible to general contractor.

b. **Nominated subcontractor**

Certain contracts permit the architect or supervising officer or the employer to reserve the right of the final selection and approval of subcontractors. In this case, the main contractor can't choose subcontractors, but will have to accept and contract with the one(s) already nominated. Sometimes nominated subcontractors (usually short as NSC) are involved in a project even before the main contractor is appointed and the project is often designed around the nominated subcontractor's features, e. g. lift shaft sizes, curtain walling features and the like. Nominated subcontractor is used usually because the work is highly specialized and can only be carried out by that firm, or because of the long-term relationship, sometimes the user specifically wants that firm. In effect the appointment of nominated subcontractors establishes a direct contractual relationship between the client and the subcontractor. But the nomination doesn't mean that main contractor have nothing to do with the nominated subcontractors, he must provide asistance and cooperation (usually provision of water, power, etc. to enable the nominated subcontractor to do his job) and take the responsibility of supervision. That is to say, he still is responsible for any subcontractors' default.

c. **Named subcontractor**

A named subcontractor, also called selected subcontractor, is employed by the main contractor in consultation with the client. The named subcontracting concept is for main subcontractor to select subcontractors from a list of would-be subcontractors provided by the consultants. So, compared with nominated subcontracting, the main contractor has more freedom in the choice of subcontractors. After the award of the subcontract, the named subcontractor is treated like any other domestic subcontractors and the employer/architect has no obligation to re-nominate.

In the past, named subcontractor and nominated subcontractor were thought to be the same in main contractor's dictionary. But the difference lies in that, with the regard to a named subcontractor, the invitation to tender and the selection of successful bidder rests with the employer's agent, while the formal tender acceptance is up to the decision of the main contractor and the resultant subcontract should act just like a domestic subcontract. But with a nominated subcontractor, the employer's agent retains some control over issues such as the selection of subcontractor, valuation of the work and the granting of extension of time.

Words, Phrases and Expressions

prominent *adj*. 杰出的，显著的，主要的（noticeable, widely known）
subcontracting *n*. 分包
subcontract *n*. 分包合同
outsourcc *vi*. 外包
approve vt 核准，批准（to agree officially to）
identifiable *adj*. 可识别的，可辨认的（recognizable）
prevalent *adj*. 普遍的，流行的（existing commonly, widely or generally）
resultant *adj*. 结果的，合成的（happening as an effect）
account for vt 说明，解决，得分（to give reasons for, to resolve）
in effect 实际上、实质上（actually, virtually）
award *n*. 裁定，判决（a grant made by a law court）
general（main）contractor：MC 总承包商
domestic subcontractor：DSC 自选分包商
nominated subcontractor：NSC 指定分包商
named subcontractor：selected subcontractor 提名分包商

疑 难 词 句

1. Subcontracting is a type of work contract that seeks to outsource certain types of work otherwise would be undertaken by general（main）contractors（sometimes

short as MC) to other independent companies (subcontractors).

译文：分包是一种工程合同形式，是将本应由总包商承担的特定工作外包给其他独立的公司，即分包商。

2. In that case, subcontracting accounts for all of the physical work done and the only responsibility of general contractor is to approve the contracts, keep the project within budget and inspect the work.

译文：在此种情况下，分包解决了所有的实际工作，总包商的唯一责任在于核准合同，保证项目按预算进行，以及对项目进行监督检查。

3. In this way, subcontractors are often contractually required (e. g. by fixed price contracts) to absorb much of the uncertainty and risk faced by the general contractor.

译文：由此，分包商通常在合同中被要求（如在固定合同价下）吸收大部分总包商所面对的不确定性和风险。

4. A subcontractor contracts with the general contractor to supply or fix any materials or goods or execute work which is part of the main contract and is fully responsible to general contractor.

译文：分包商与总包商签订合同供应或修理原材料或物品，或执行总包商的一部分任务，并对总包商负全责。

5. In effect the appointment of nominated subcontractors establishes a direct contractual relationship between the client and the subcontractor.

译文：实质上，对指定分包商的任命在业主与分包商之间建立起了直接的合同关系。

6. After the award of the subcontract, the named subcontractor is treated like any other domestic subcontractors and the employer/architect has no obligation to re-nominate.

译文：一旦达成了分包合同，提名分包商将像其他任何自选分包商一样，雇主/建筑师无权重新指定。

7. But the difference lies in that, with the regard to a named subcontractor, the invitation to tender and the selection of successful bidder rests with the employer's agent, while the formal tender acceptance is up to the decision of the main contractor and the resultant subcontract should act just like a domestic subcontract.

译文：但是区别在于对于提名分包商来说，投标邀请和选择中标者的权力在于雇主的代理人，而正式的接受投标则取决于总包商，由此签订的分包合同应视同指定分包合同一样看待。

11.2 Domestic subcontractor

11.2.1 Selection of domestic subcontractors

Contractors require domestic subcontractors that are fully qualified, or else they may become another risk. Because the contractor-subcontractor relationships is project-based, the selection of subcontractors is one of the key issues for main contractors who pursue good performance and reputation. Since main contractor has more freedom in the selection of subcontractors in domestic subcontracting, we will focus on domestic subcontractors here.

The subcontractors are required to have adequate skills and resource to execute work effectively at an agreed price and quality arrived at in the subcontract. Usually, contractor has different ways to select subcontractors, some of which are listed below:

1. Single source subcontractor: it happens when main contractor has only one subcontractor to choose for a specific trade, which has developed good relationship with main contractor through previous cooperation and is fully integrated within main contractor's business;

2. Specialist subcontractor: when main contractor knows the subcontractors who could deliver specialist tasks and services and uses them whenever required;

3. Long list of subcontractors: when main contractor has selected a number of subcontractors for each trade work, and got them compete with each other through tender for getting the lowest price; and

4. open tender: the main contractor may use this type of tendering when he does not have a long list of subcontractors and looks for new partners in a traditional way, or when he already has potential choice but still wants to test the market or to put pressure on subcontractors for lower price or more risk sharing.

A general practice in the selection of subcontractor emphasizes tender price. But in spite of the advantages such as time saving, it may easily ignore competent subcontractors. A lower tender price is not equivalent to better competency nor does it guarantee good performance. Therefore, other criterions should be taken into consideration to measure a qualified subcontractor. The criterions may require the subcontractors to:

- Produce project schedule
- Produce quality work
- Employ skilled staff
- Cultivate goodwill between parties
- Be able to allocate resources

There are elaborate procedures for the selection of nominated subcontractors. But for domestic subcontractor selection, some people say that there is no well designed procedure, while others argue that contractors in practice do use a series of criterions unknown to outside world. According to CIB World Building Congress, the selection should be made under the principles of:

- Clear procedures that ensure fair and transparent competition among all subcontractors;

• The tendering process that shortlists systematically from a number of qualified candidates;

• Formal and recognizable contracts bodies that should be used where they are available; and

• A commitment to team work from all parties.

11.2.2 Relationship between the main contractor and subcontractors

Contractor, as the party who signs general contract with the client, should be totally responsible to client both legally and financially. Actually, according to Sub-Clause 4.1 of FIDIC Conditions of Contract for Works of Civil Engineering Construction (the Red Book), fourth edition, the Contractor may not subcontract the whole of the Works. Except where otherwise provided by the contract, the Contractor shall not subcontract any part of the Works without the prior consent of the Engineer). And as the party who outsources part of his job to third parties, contractor should undertake the job of administrating subcontractors according to subcontract agreement. In subcontracting, the main contractor is supposed to support and provide enabling environment for subcontractors in order to fulfill their job. Most importantly, he is also supposed to take care of the subcontractors' job by supervising their design, the quality and progress of the project, and take full responsibility if there is any default by the latter. From the point of subcontractors, they will have to accept the co-ordination arrangements and scheduling of the main contractor, and to assume all the responsibilities written in the subcontracting agreement.

Any default from main contractor to subcontractor, the latter's economic loss should be paid by the former. Similarly, subcontractors should compensate for the loss of the main contractor, if the latter is fined or sanctioned for any default of the former, as is provided in Sub-Clause 4.4 of Conditions of Subcontract for Work of Civil Engineering Construction (1st Edition, 1994): If the Subcontractor commits any breaches of the Subcontract, he shall indemnify the Contractor against any damages for which the Contractor becomes liable under the Main Contract, as a result of such breaches. In such event the Contractor may, without prejudice to any other method of recovery, deduct such damages from recoveries otherwise becoming due to the Subcontractor.

The client is not involved in the selection of domestic subcontractor; neither will he make the payment directly to subcontractors. The payment is made by the main contractor, who gets the project payment from the client before subcontracting and pays the subcontractors their part according to agreement and their performance.

Words, Phrases and Expressions

reputation *n.* 名声，名誉，声望，名望 (fame)

arrive at *vi.* 获得（结果），达到（目的）(to reach，come to)
specific trade 特定任务
specialist subcontractor 专业分包商
open tender 公开招标
goodwill *n.* 商誉，信誉（credibility）
co-ordination *n.* 协调，统筹（plan as a whole）
scheduling *n.* 日程安排（setting an order and time for planned events）
indemnify *vi.* 赔偿，补偿，保护（to compensate for，to make amends for ）
be liable for 对……应负责任（be responsible for）
recovery *n.* 回收，财产收回，追偿（reclaim）
without prejudice (to)【律】不使（合法权利）受到损害，无损于，无害于，不影响
deduct *vi.* 扣除，减去（to take away or subtract）
CIB World Building Congress CIB 国际建筑研讨会
(CIB is the acronym of the abbreviated French (former) name: "Conseil International du Bâtiment" (in English this is: International Council for Building). In the course of 1998, the abbreviation has been kept but the full name changed into: International Council for Research and Innovation in Building and Construction.

疑 难 词 句

1. Actually, according to Sub-Clause 4. 1 of FIDIC Conditions of Contract for Works of Civil Engineering Construction (the Red Book), fourth edition, the Contractor may not subcontract the whole of the Works. Except where otherwise provided by the contract, the Contractor shall not subcontract any part of the Works without the prior consent of the Engineer).

译文：实际上，根据 FIDIC《土木工程施工合同条件》（简称红皮书）第四版，承包人不得将整个合同全部分包出去，除非合同另有规定，承包人不应在未得到工程师的同意前将合同的任何部分分包出去。

2. from the point of subcontractors, they will have to accept the co-ordination arrangements and scheduling of the main contractor, and to assume all the responsibilities written in the subcontracting agreement

译文：从分包商的角度讲，他们必须接受总承包商的统筹及时间安排，并承担分包合同中规定的所有责任。

3. If the Subcontractor commits any breaches of the Subcontract, he shall indemnify the Contractor against any damages for which the Contractor becomes liable under the Main Contract, as a result of such breaches.

译文：如果分包商就分包合同违约，他应就总承包商在总包合同项下因违约承担的任何损失进行赔偿。

4. In such event the Contractor may, without prejudice to any other method of

recovery, deduct such damages from recoveries otherwise becoming due to the Subcontractor.

译文：这种情况下，承包商可以在不影响其他追偿方式的前提下，从本应付给分包商的款项中将该损失予以扣减。

11.3 Nominated Subcontracting

We have defined nominated subcontractor in the earlier part of this chapter, thus the process by which the employer nominates, selects or approves who will perform a subcontract or specialist trade is called nominated subcontracting. The NSC then enters into a subcontract with the Main Contractor ("MC"). It is a means for the Employer to retain some control over the specialist contractor or supplier without necessarily becoming directly involved in detailed contractual arrangements with the specialist.

11.3.1 Why Nominate

Nomination is used because there are benefits for the Employer in using the system. The key benefit for the Employer is that he can still control the selection of subcontractors but doesn't necessarily become directly involved in detailed contractual arrangements with them. Further, the Employer can, if he wishes, control the terms of the subcontract, including the price and scope of nominated subcontractors' work. Another benefit is obvious for some works in which specialist subcontract work requires a longer lead time than the construction program would allow. Such work must be started even before a main contractor is decided upon. So the reasonable way would be nomination.

11.3.2 Difference between nominated subcontractors and domestic subcontractors

Apparently, NSC is different from DSC in many ways:

a. the selection of subcontractors: as we already know, DSC is decided by main contractor, but the selection of NSC is up to employer or engineer. During the process, the main contractor is entitled to raise reasonable objection, but only with adequate and reasonable justification.

b. the nature of job: usually, DSC comes into the project to undertake job which is part of the main contract. While comparatively, NSCs are involved to finish what the main contractor is not competent in and what is not included in the main contract.

c. the payment to subcontractors: the payment to DSCs is a part of the total contract value and is paid by main contractor in accordance with the subcontract. In contrast, the payment to NSCs, also provided by main contractor who gets the amount previously from the employer, is paid out of the employer's provisional sum, the amount set aside specifically in order to protect the benefit of main contractor. And the main contractor can only charge from NSCs the administration fee arrived at in the subcontract agreement.

d. NSCs must guarantee the interests of the main contractor. Despite the fact that NSCs are appointed by employer, once a main contractor has accepted the nomination proposal, he becomes responsible for that subcontractor, just as he would for his own DSCs. Any misdeeds by NSCs will directly harm the benefit of the main contractor. So, the main contractor would require some guarantees from NSCs just in case. In nominated subcontract, the NSCs will have to make commitments as stated in Sub-Clause 59. 2 of FIDIC Conditions of Contract for Works of Civil Engineering Construction:

- that in respect of the work, goods, materials, plant or services the subject of the Subcontract, the nominated Subcontractor will undertake towards the Contractor such obligations and liabilities as will enable the Contractor to discharge his own obligations and liabilities toward the Employer under the terms of the Contract and will save harmless and indemnify the Contractor from and against the same and from all claims, proceedings, damages, costs, charges and expenses whatsoever arising out of or in connection therewith or arising out of or in connection with any failure to perform such obligations or to fulfill such liabilities, and
- that the nominated Sub-Contractor will save harmless and indemnify the Contractor from and against any negligence by the nominated Sub-Contractor, his agents, workmen and servants and from and against any misuse by him or them of any Constructional Plant or Temporary Works provided by the Contractor for the purposes of the Contract and from all claims as aforesaid.

NSCs' refusal to take such commitment will give the main contractor reason to object to the nomination.

e. Handling method when breaches arise: the engineer has the right and obligation at a reasonable time to replace the NSC if there is any breaches by any NSCs. The failure to do so will incur the suspension of work or extended period requirement from the main contractor.

Words, Phrases and Expressions

lead time *n.* 完成某项活动所需的时间

competent *adj.* 能干的，胜任的 (capable, suitable or sufficient for the purpose)

provisional sum 暂定金额有时也叫待定金额或备用金

set aside *vi.* 留出 (to reserve for a special purpose)

administration fee 管理费

commitment *n.* 承诺，约定 (an obligation, promise)

aforesaid *adj.* 前述的，该（aforementioned）
justification *n.* 理由（reason）
in respect of 关于，就……而言（on the terms of）
negligence *n.* 疏忽，粗心大意，忽视（inattention）

疑 难 词 句

1. The process by which the employer nominates, selects or approves who will perform a subcontract or specialist trade is called nominated subcontracting.

译文：由雇主指定、选择、同意由谁执行分包合同或专业工程的程序叫做指定分包。

2. During the process, the main contractor is entitled to raise reasonable objection, but only with adequate and reasonable justification.

译文：在此过程中，总承包商有权提出合理的反对意见，但必须要有充分合理的理由。

3. In contrast, the payment to NSCs, also provided by main contractor who gets the amount previously from the employer, is paid out of the employer's provisional sum, the amount set aside specifically in order to protect the benefit of main contractor.

译文：相比之下，指定分包商的工程款虽然也是由业主先行支付给总承包商，但业主是在"暂定金额"项下进行开支，即另外准备的款项中开支，以不损害总承包商的利益。

4. that in respect of the work, goods, materials, plant or services the subject of the Subcontract, the nominated Subcontractor will undertake towards the Contractor, such obligations and liabilities as will enable the Contractor to discharge his own obligations and liabilities toward the Employer under the terms of the Contract and will save harmless and indemnify the Contractor from and against the same and from all claims, proceedings, damages, costs, charges and expenses whatsoever arising out of or in connection therewith or arising out of or in connection with any failure to perform such obligations or to fulfill such liabilities.

译文：就工程、货物、材料或服务，转包合同当事人，即指定分包人将对承包人承担本合同所规定的承包人应对业主承担的同样的义务和责任，并不让承包人再履行此义务和责任，且不让其受因履行此种义务和责任而产生或与之有关的，或因不履行此种义务或责任而产生或与之有关的一切索赔、诉讼、损害赔偿金、诉讼费、开支和费用的损害。

5. that the nominated Sub-Contractor will save harmless and indemnify the Contractor from and against any negligence by the nominated Sub-Contractor, his agents, workmen and servants and from and against any misuse by him or them of any Constructional Plant or Temporary Works provided by the Contractor for the purposes of the Contract and from all claims as aforesaid.

译文：指定分包人应使承包人不因指定分包人、其代理人、工人和雇员的过失而受任何损害，不因分包人、其代理人，工人和雇员不当使用承包人按合同所提供的工程设备或临建工程而受损害，以及不因上述所有的索赔而受损害。

11.4 Payment to subcontractors

Performance by subcontractors in accordance with the provisions of the contract shall entitle them to payment from the main contractor except when the subcontract agreement regulates otherwise, like perhaps in nomination. In practice, when time comes that subcontractors should be paid, main contractors will, out of their own benefit, usually focus on issues such as but not limited to the following:

a. they make partial payment monthly to subcontractors, computed on the basis of the prices set forth between the two parties and the actual quantity of work performed, with less retainage and less aggregate of previous payments. But such partial payments shall not become due until certain days (say 10) after the main contractors get paid for such work from the employer. In practice, the main contractors, for purpose of protecting their own interest, sometimes include in their subcontracts conditional payment clauses. As an example, they may include clause such as payment from the main contractor to the subcontractors on payment from the owner to the main contractor. These clauses, depending on their languages, are commonly known as "pay if paid" or "pay when paid" clause. If contractors receive payment from Owner for less than the full value of materials delivered to the site but not yet incorporated into the Work, the amount due Subcontractors on account of such material delivered to the site shall be proportionately reduced. No partial payment to Subcontractors shall operate as approval or acceptance of work furnished hereafter.

b. main contractors will make final payment to Subcontractor of the balance due to it within a reasonable time after full payment of such work from employer to Contractor. But the final payment is made only upon complete performance of the Subcontract agreement by Subcontractor, and final approval and acceptance of Subcontractor's work by employer.

c. If at any time prior to final payment Owner reduces the retainage withheld from Contractor, Contractor may, in its sole discretion, reduce retainage withheld from Subcontractor, but usually with the consent of Subcontractor's surety.

d. if there is any sum owed by Subcontractor to Contractor, Contractor can make the deduction from any amounts due or become due Subcontractor. Similarly, Contractor may also set off any sums owned by Subcontractor and/or its affiliates.

e. any loss, damage or extra expense to contractor due to any breach by Subcontractor of any provision or obligation of the Subcontract shall entitle contractor the right to retain out of payment due subcontractor an amount sufficient to completely protect contractor's benefit until the situation has been satisfactorily fixed by Subcontractor.

Words, Phrases and Expressions

partial payment 部分付款
retainage *n.* 留置金（retention payments）
aggregate *n.* 合计，总计（sum total）
conditional payment clauses 有条件付款条款
hereafter *adv.* 今后，从此以后（at some time in the future）
balance *n.* 余额（money or something else which remains or is left over）
affiliate *n.* 附属机构（subsidiary bodies）
pay if paid 只有已经得到付款时才必须支付相应款项
pay when paid 得到付款时支付相应款项
incorporated into 纳入

疑 难 词 句

1. Payment from the main contractor to the subcontractors on payment from the owner to the main contractor

译文：只有雇主向总包商付款，总包商才向分包商付款。

2. If contractors receive payment from Owner for less than the full value of materials delivered to the site but not yet incorporated into the Work, the amount due Subcontractors on account of such material delivered to the site shall be proportionately reduced.

译文：如果总包商从雇主那里得到款项少于已运到工地但尚未纳入工程的材料总额，则（总包商）可以就该材料而应向分包商支付的款项予以相应比例的扣除。

3. But the final payment is made only upon complete performance of the Subcontract agreement by Subcontractor, and final approval and acceptance of Subcontractor's work by employer.

译文：但是最终的付款只有在分包商全部完成分包合同，并获得雇主对其工作的认可和接受之后才能予以支付。

4. If at any time prior to final payment Owner reduces the retainage withheld from Contractor, Contractor may, in its sole discretion, reduce retainage withheld from Subcontractor, but usually with the consent of Subcontractor's surety.

译文：如果在最终付款之前雇主减少了其对总包商的留置金，则总包商也可以自行决定减少其对分包商的留置金，但条件是分包商同意提供担保。

5. If there is any sum owed by Subcontractor to Contractor, Contractor can make the deduction from any amounts due or become due Subcontractor.

译文：如果分包商对总包商有欠款，总包商可以从其对分包商的欠款中进行抵扣。

6. any loss, damage or extra expenses to contractor due to any breach by Subcontractor of any provision or obligation of the Subcontract shall entitle contractor the right to

retain out of payment due subcontractor an amount sufficient to completely protect contractor's benefit until the situation has been satisfactorily fixed by Subcontractor.

译文：如果分包商就合同条款及责任违约从而造成总包商的任何损失及额外的费用，那么直到分包商圆满解决问题之前，总包商都有权从对分包商的欠款中留置出足够的足以保护总包商利益的金额。

Chapter 12 Suspension and Termination of Contract

Situations may occur during contract performance that causes the client to order a suspension or termination of work. In this chapter, we will discuss the circumstances in which the contract work may be lawfully stopped, either temporarily or permanently. Whatever it is, remedies should be made for the suspension of work or termination of the contract for breach.

12.1 Suspension of Work

Construction contracts often permit the Employer and sometimes the Contractor to order a temporary suspension of work under certain circumstances. A suspension of work under a construction or architect-engineer contract may be ordered by the contracting officer for a reasonable period of time. If the suspension is unreasonable, the Contractor may submit a written claim for increases in the cost of performance, excluding profit.

12.1.1 Suspension by Employer

Under any construction agreements, the Employer may, with or without stating any reasons, suspend execution of the tasks under the contract or any part thereof, which are respectively called suspension for cause and suspension for convenience. Suspension shall take effect on the day the Contractor receives notification by registered letter with acknowledgment of receipt or equivalent, or at a later date where the notification so provides.

The Employer may at any time instruct the Contractor to suspend progress of part or all of the Works. During such suspension, the Contractor shall protect, store and secure such part of the Works against any deterioration, loss or damage.

Should the Employer require a temporary stoppage for cause other than the Contractor's fault, it is only fair and equitable that the Contractor be reimbursed for all extra costs actually incurred including profit on the increased costs. If completion is or will be delayed due to suspension, the Contractor will usually be entitled to an extension of time. However, contracts vary significantly concerning the precise extent of that entitlement. If after the period of suspension the work is resumed, the Contractor may also want to address the scenario of rising costs during the period of suspension.

But if the cause of suspension is proved to be the fault of the Contractor's, the Contractor's right to ask for reimbursement is deprived. Under Sub-Clause 8.9 of the FIDIC Conditions of Contract for Construction, the Contractor shall not be entitled to an extension of time for, or to payment of the cost incurred in, making good the consequences of the Contractor's faulty design, workmanship or materials, or of the Contractor's failure to protect, store or secure such part or the Works against any deterioration, loss or damage.

Suspension will usually impose serious commercial and financial influence upon Contractor. Therefore, construction contracts often allow a contractor to terminate the contract instead of letting it run indefinitely, if suspension continues beyond a stipulated period of time. Normally, termination will require formal notice from the contractor and there may be a complex sequence involving a preliminary request to the employer for permission to proceed and then, if the request is not granted, a contractor's notice treating suspension either as an omission of part of the works or as grounds for termination. For example, according to Sub-Clause 8.11 of the FIDIC Conditions of Contract for Construction, if the suspension has continued for more than 84 days, the Contractor may request the Employer's permission to proceed. If the Employer does not give permission within 28 days after being requested to do so, the Contractor may, by giving notice to the Employer, treat the suspension as an omission of the affected part of the Works.

Words, Phrases and Expressions

termination *n.* 终止，终结；结局，结束 (stop, end)
execution *n.* 实行，履行，执行；贯彻 (implementation, carrying out)
thereof *adv.* 它的；由此 (its)
registered letter *n.* 挂号信
suspension for cause 有原因的暂停
suspension for convenience 随意暂停
notification *n.* 通告书，通知单；报告书 (announcement)
acknowledgment *n.* 承认 (recognition, confession)
receipt *n.* 接受，接收；收条，收据 (reception, proof of payment, note)
equivalent *adj.* 相当的，相同的，同等的 (alike, the same)
deterioration *n.* 恶化；变质；退化 (worsening, decline)
equitable *adj.* 公平的，公正的，合理的 (fair, reasonable, impartial)
scenario *n.* 情况 (setting, situation)
reimbursement *n.* 偿还，赔偿，补偿 (compensation, repayment)
deprive *vi.* 剥夺 (take away, rob)
indefinitely *adv.* 无定限的，无限期的 (for an indefinite period, for ever)

stipulate *vi.* 约定，订定；规定，订明（demand，specify）
preliminary *adj.* 预备的；初步的，初级的（initial，introductory）

疑 难 词 句

1. Under any construction agreements, the Employer may, with or without stating any reasons, suspend execution of the tasks under the contract or any part thereof, which are respectively called suspension for cause and suspension for convenience. Suspension shall take effect on the day the Contractor receives notification by registered letter with acknowledgment of receipt or equivalent, or at a later date where the notification so provides.

译文：在任何施工协议中，雇主都可以在说明或不说明原因的情况下暂停实施合同项下全部或部分任务，这分别被称作有原因的暂停和随意暂停。暂停于承包商收到以挂号信形式寄出的通知并承认收到当天生效，或在通知规定的他日生效。

2. The Employer may at any time instruct the Contractor to suspend progress of part or all of the Works. During such suspension, the Contractor shall protect, store and secure such part or the Works against any deterioration, loss or damage.

译文：雇主可以随时指示承包商暂停工程某一部分或全部的施工。在暂停期间，承包商应保护、妥善保管该部分或全部工程不致产生任何变质、损失或损害。

3. Should the Employer require a temporary stoppage for cause other than the Contractor's fault, it is only fair and equitable that the Contractor be reimbursed for all extra costs actually incurred including profit on the increased costs.

译文：如果雇主因承包商违约之外的其他原因要求暂时停止，那么只有承包商的所有额外成本包括成本提高后的利润都得到补偿才算公正合理。

4. Under Sub-Clause 8.9 of the FIDIC Conditions of Contract for Construction, the Contractor shall not be entitled to an extension of time for, or to payment of the cost incurred in, making good the consequences of the Contractor's faulty design, workmanship or materials, or of the Contractor's failure to protect, store or secure such part or the Works against any deterioration, loss or damage.

译文：FIDIC《施工合同条件》第8.9款规定，承包商应无权得到为弥补因承包商有缺陷的设计、工艺或材料，或因承包商未能保护、保管或保证安全的后果，带来的延长期和招致费用的支付。

5. Normally, termination will require formal notice from the Contractor and there may be a complex sequence involving a preliminary request to the Employer for permission to proceed and then, if the request is not granted, a Contractor's notice treating suspension either as an omission of part of the works or as grounds for termination.

译文：通常，终止合同要有承包商开出的正式通知书，这将产生复杂的程序，包括承包商先向雇主请求允许其继续工作，继而，如果未获批准，承包商要发出通知书，将暂停当做是删除一部分项目或作为项目终止的依据。

12.1.2 Suspension by Contractor

A Contractor should also ensure that they have an equal right to suspend work as the Employer does. One should bear in mind that notwithstanding what the contract says, the Contractor has a statutory right to suspend performance of services for non-payment or for some other serious but remediable breach of contract. However, suspension is by no means universally allowed, and incorrect or unlawful suspension may amount to a repudiation of contract with potentially serious consequences. Therefore, the decision of suspension should be made with deliberation.

Some contracts provide an express right to suspend works in case of nonpayment. For example, Sub-Clause **16.1** of FIDIC Conditions of Contract for Construction [***Contractor's Entitlement to Suspend Work***] regulates that if the Employer fails to comply with Sub-Clause **2.4** [Employer's Financial Arrangements] or Sub-Clause **14.7** [Payments], the Contractor may, after giving not less than **21** days' notice to the Employer, suspend work (or reduce the rate of work) unless and until the Contractor has received the Payment Certificate, reasonable evidence or payment, as the case may be and as described in the contract.

Many other contracts, however, do not directly provide a contractor the right to suspend his work for nonpayment, but keep silent regarding the right or seem to deny that right instead. In that case, the contractor can still seek justice and protect his own interest by turning to general principles of law governing the contract. Some civil law jurisdictions, such as France, provide a legal right for a contracting party to suspend where the other party is in breach of contract. But still, in common law jurisdictions such as England and Hong Kong, there are no generally recognized right for a contractor to suspend work for nonpayment.

If after a fixed period of suspension, the contractor still doesn't get paid, he may want to terminate the work. But if he does get the payment or evidence, he shall resume the work suspended. For example, according to Sub-Clause 16.1 of the FIDIC Conditions of Contract for Construction, if the Contractor subsequently receives such evidence or payment before giving a notice of termination, the Contractor shall resume normal working as soon as is reasonably practicable.

In the clause of suspension, the contractor must make sure that he is treated and indemnified fairly, either in the form of extension of time or in financial compensation. Sub-Clause **16.1** of the FIDIC Conditions of Contract for Construction says that if the Contractor suffers delay and/or incurs Cost as a result of suspending work (or reducing the rate of work) in accordance with this Sub-Clause, the Contractor shall give notice to the Employer and shall be entitled subject to Sub-Clause **20.1** [***Contractor's Claims***] to:

a. an extension of time for any such delay, if completion is or will be delayed, under Sub-Clause 8.4 [*Extension of Time for Completion*], and

b. payment of any such Cost plus reasonable profit, which shall be added to the Contract Price. After receiving this notice, the Employer shall proceed in accordance with Sub-Clause 3.5 [*Determinations*] to agree or determine these matters.

Words, Phrases and Expressions

bear in mind 记住 (to remember)
notwithstanding *adv.* 虽然，尽管……仍…… (although, even though)
remediable *adj.* 可挽回的，可补救的 (reparable, retrieval)
repudiation *n.* 推翻，批判，驳斥 (denial, refutation)
amount to 等同于 (be equivalent to)
deliberation *n.* 深思熟虑，慎重 (consideration, thoughtfulness)
express *adj.* 直接的 (direct)
Payment Certificate 支付证明
civil law 民法
jurisdiction *n.* 裁判权；司法；司法权 (judicial power)
be subject to 以……为条件［转移］的，必须得到……的

疑 难 词 句

1. One should bear in mind that notwithstanding what the contract says, the Contractor has a statutory right to suspend performance of services for non-payment or for some other serious but remediable breach of contract.

译文：应该记住的是，无论合同条款如何规定，承包商都有法定的权利因未得到付款或其他严重的但可挽回的违约而暂停提供服务。

2. For example, clause 16.1 of the FIDIC *conditions of contract for construction* [*Contractor's Entitlement to Suspend Work*] regulates that if the Employer fails to comply with Sub-Clause 2.4 [Employer's Financial Arrangements] or Sub-Clause 14.7 [*Payments*], the Contractor may, after giving not less than 21 days' notice to the Employer, suspend work (or reduce the rate of work) unless and until the Contractor has received the Payment Certificate, reasonable evidence or payment, as the case may be and as described in the contract.

译文：比如，FIDIC《施工合同条件》第 16.1 款（承包商暂停项目的权利）规定，如果雇主没能遵守第 2.4 款（雇主的财务安排条款）或者第 14.7 款（付款条款）规定，承包商可以提前 21 天以上通知雇主暂停工作（或降低工作速度），除非并且直到承包商收到了支付证书，合理的证明或支付（根据情况与合同的规定）。发出通知（至少 21 天的通知）之后暂停工作。

3. Clause 16.1 of the FIDIC *Conditions of Contract for Construction* says that if the

Contractor suffers delay and/or incurs Cost as a result of suspending work (or reducing the rate of work) in accordance with this Sub-Clause, the Contractor shall give notice to the Employer and shall be entitled subject to Sub-Clause 20. 1 [*Contractor's Claims*] to:

译文：FIDIC《施工合同条件》第 16. 1 款规定，如果承包商根据本款规定暂停工作或降低工作速度而造成拖期和（或）导致发生费用，则承包商应通知雇主，并根据第 20. 1 款（承包商的索赔）承包商有以下权利：

12. 2 Termination of Work

Generally, the most welcome termination occurs by final acceptance and contract closure, i. e. when the entire performance of the contract has been met and all the parties are satisfied. However, the contract may also be terminated under other circumstances, which is the main concern in this chapter. In some cases, the employer may want to terminate the contract when the contractor is unable to complete the job, or when he encounters serious financial difficulties on the project and can't complete it. Sometimes, the employer himself may run into financial difficulties and have to terminate the contract. With the above causes for termination being reasonable, this situation is also called termination for cause. In other cases, the employer may terminate the contract for his own convenience without any apparent reasons, or when he runs into frustration and has to give up the construction, which is called termination for convenience. Termination for cause is a right of a contractor and employer to completely or partially terminate performance of the contract under the provisions of a termination clause, whereas termination for convenience is a clause allowing one (or more) parties to a contract to terminate that contract at any time without reason. Under the term of termination for convenience, the contractor is generally entitled to a negotiated settlement for an equitable recovery of costs and losses incurred.

Construction contract entitles each party a termination right. But each party has to bear in mind that if the claimed default is found to be improper and unjustified, he may be liable for damages for improper termination and have to face significant damage claims including profits loss of the terminated contractor. Furthermore, even if the decision of termination is reasonable, the extra time and cost needed makes termination by employer an issue that requires careful consideration. For example, from the perspective of the employer, the decision of termination will result in expensive litigation or arbitration. The contractor and its subcontractors will probably file liens against the property, possibly creating problems for the employer with the lender, which may pull the employer into a lengthy and intricate dispute resolution process. In addition, hiring a new contractor to take over the work will be difficult and expensive. The architect will probably charge for the additional time required because of the change of contractors. Therefore, the decision to terminate the contractor is not one to be taken lightly.

Words, Phrases and Expressions

termination for cause 有原因地终止合同
termination for convenience 任意解除合同
termination for default 因违约终止合同
be liable for 对什么负责 (be responsible for)
lien *n.* 留置权
intricate *adj.* 缠结的，错综的，复杂的 (complex, complicated)
lightly *adv.* 掉以轻心 (carelessly)

疑 难 词 句

1. Termination for cause is a right of a contractor and employer to completely or partially terminate performance of the contract under the provisions of a termination clause.

译文：有原因地终止合同是指承包商和雇主有权利在终止条款下全部或部分地终止合同的执行。

2. Under the term of termination for convenience, the contractor is generally entitled to a negotiated settlement for an equitable recovery of costs and losses incurred.

译文：在任意解除合同项下，承包商通常有权通过协商解决，以弥补因终止合同而导致的成本以及损失。

3. But each party has to bear in mind that if the claimed default is found to be improper and unjustified, he may be liable for damages for improper termination and have to face significant damage claims including profits loss of the terminated contractor. Furthermore, even if the decision of termination is reasonable, the extra time and cost needed makes termination by employer an issue that requires careful consideration.

译文：但是每一方都必须牢记，如果所提出的违约索赔不恰当或无正当理由，他可能对该不当的终止而造成的损失负责，从而不得不面对巨额的损失索赔，包括被终止的承包商的利润损失。进一步说，即使终止的决定合情合理，所需的额外时间和成本也应使雇主慎用终止。

12.3 Termination by Employer

The employer of the project can terminate the contract either for default from the contractor or for his own convenience. Termination for default, is a clause which gives an employer the right to unilaterally terminate the contract if the contractor fails to perform according to the specified terms. The right to terminate a contract depends on the nature and the consequences of the contractor's breach. The breach must either be of a fundamental term of the contract, or alternatively the consequences of the breach must be such that they substantially deprive the employer of the entire benefit intended by the contract.

Here is a typical clause for termination by employer. The Employer may at any time terminate this Agreement for cause by giving notice to the Contractor, in writing, to that effect not less than thirty days prior to the effective date of termination specified in this notice. Such notice shall be deemed given if delivered or mailed to the last known address of the contractor.

12. 3. 1 The Event Incurring Termination for Default

The employer may declare the contractor to be in default under the agreement and may terminate the contractor's right to proceed thereunder if the latter:

- becomes bankrupt or otherwise gets into serious financial difficulties.
- is unable to complete the work in accordance with the term of the contract.
- does not execute the work in a timely fashion.

The Sub-Clause 15. 2 of FIDIC Conditions of Contract for Construction defines the situation in detail, in which the employer shall be entitled to terminate the contract if the contractor:

- fails to comply with performance security.
- abandons the Works or otherwise plainly demonstrates the intention not to continue the performance of his obligations under the contract.
- without reasonable excuse fails to:

(ⅰ) proceed with the Works in accordance with Clause 8 [*Commencement, Delays and Suspension*], or

(ⅱ) comply with a notice issued under Sub-Clause 7. 5 [*Rejection*] or Sub-Clause 7. 6 [*Remedial Work*] within 28 days after receiving it.

- subcontracts the whole of the Works or assigns the contract without the required agreement.
- becomes bankrupt or insolvent, goes into liquidation, has a receiving or administration order made against him, compounds with his creditors, or carries on business under a receiver, trustee or manager for the benefit of his creditors, or if any act is done or event occurs which (under applicable Laws) has a similar effect to any of these effects or events, or
- gives or offers to give (directly or indirectly) to any person any bribe, gift, gratuity, commission or other things of value, as an inducement or reward:

(ⅰ) for doing or forbearing to do any action in relation to the contract, or

(ⅱ) for showing or forbearing to show favor or disfavor to any person in relation to the contract.

Or if any of the Contractor's personnel, agents or subcontractors gives or offers to give (directly or indirectly) to any person any such inducement or reward as is described in Sub-Cause f. However, lawful inducement or rewards to Contractor's personnel shall not entitle termination.

In any of these events or circumstances, the employer may, upon giving 14 days notice to the contractor, terminate the contract and expel the contractor from the site. However, in the sub-paragraph e or f, the employer may by notice terminate the contract immediately.

A termination for cause article is one of those clauses in a construction contract that the employer will hold on to and one that the contractor has to live with. For some contractors, it is like a time bomb. But for contractors with good financial standing and the ability to perform the work according to specification and schedule, a termination for cause article should present little risk.

Words, Phrases and Expressions

to that effect 有这样/那样的意思或内容

thereunder *adv.* 在其下；在那一项目［条款］下

liquidation *n.* 清理，清算；（债务的）清偿（insolvency, bankruptcy）

gratuity *n.* 赏金，小费（bonus, tips）

commission *n.* 手续费，佣金（brokerage）

forbear *vi.* 忍耐，克制，避免（refrain, withhold）

expel *vi.* 赶出；驱逐；开除（drive out）

financial standing 财务状况

疑难词句

1. Termination for default, is a clause which gives an Employer the right to unilaterally terminate the contract if the Contractor fails to perform according to the specified terms.

译文：因违约而终止合同，这一条款赋予雇主因承包商未能按规定条款履约时单方面终止合同的权利。

2. The breach must either be of a fundamental term of the contract, or alternatively the consequences of the breach must be such that they substantially deprive the employer of the entire benefit intended by the contract.

译文：该违约必须是对合同基础条款的违约，或者违约的结果是实质上剥夺了雇主应从合同中获得的好处。

3. The Employer may at any time terminate this Agreement for cause by giving notice to the Contractor, in writing, to that effect not less than thirty days prior to the effective date of termination specified in this notice.

译文：雇主可以在任何时间向承包商发出书面通知，以某种缘由终止该协议，该通知要在其规定的终止生效日至少30天之前发出。

4. (The Contractor) becomes bankrupt or insolvent, goes into liquidation, has a receiving or administration order made against him, compounds with his creditors, or carries on business under a receiver, trustee or manager for the benefit of his creditors, or if any act is done or event occurs which (under applicable Laws) has a similar effect to any of these effects or events.

译文：承包商破产或无力偿还债务，或停业清理，或已由法院委派其破产财产管理人或遗产管理人，或为其债权人的利益与债权人达成有关协议，或在财产管理人、财产委托人或财务管理人的监督下营业，或承包商所采取的任何行动或发生的任何事件（根据有关适用的法律）具有与前述行动或事件相似的效果。

5. A termination for cause article is one of those clauses in a construction contract that the employer will hold on to and one that the contractor has to live with.

译文：在施工合同条款中，有原因地终止合同这一条款雇主必须紧紧抓住，而承包商也不得不面对。

12.3.2 Steps of Termination for Cause Taken by Employer

a. **written certification from the architect**

To terminate a contract due to default by the contractor, the employer must obtain the architect's written certification that sufficient cause exists to justify the employer's proposed action. For example, FIDIC 4th Edition requires a certificate to be issued by the engineer stating that the contractor:

• has repudiated the contract, or

• without reasonable excuse has failed to commence or proceed with the works following a 28-day notice, or

• has failed to comply with a specific notice or instruction concerning rejection and/or removal of improper work, materials or plant, or

• despite previous warning from the engineer, is neglecting to comply with any of the contractor's obligations, or

• has contravened the provisions of the contract related to sub-contracting, in which case, upon giving 14 days' notice to the contractor, the employer may enter the site and terminate the contract.

The architect who executes such a certification should make a positive independent determination that any causes are true and can be proved by documentary, photographic, or other convincing evidence.

b. **termination notice issuance from the employer**

Once the default of the contractor is testified, the employer must give the contractor and the contractor's surety, if any, certain days' written notice prior to the date of termination (e. g. seven days in a standard AIA form of agreement and fourteen days in FIDIC 4th Edition).

However, if, during the notice period, the contractor makes a successful effort to correct its misdeeds and rehabilitate its position, the employer could choose to set aside the termination option. Economically, this is often the best solution to the problem for both parties, although the employer might have lost confidence in the contractor. Often this is virtually impossible for the contractor to do within the short period. If the breaches or deficiencies are not cured by contractor within the time limit, the termination will become effective at the end of the notice period. At that time, the employer should issue a final notice of termination. In construction contract, prior written notice provisions should be specified and the obligations of both parties on termination set out clearly, i. e. the employer's obligation to pay a contractor and the contractor's obligations to clear the site, etc. Usually, as proof that the contractor has received each notice, all of the notices (written notice prior to the date of termination and the final notice of termination) should be sent by registered mail or personal delivery.

c. **works after termination**

After the contract has been duly terminated, the contractor shall then leave the site and deliver to the engineer all of the contractor's documents and other design documents made by or for him for latter use. The employer is entitled to take over the site and all materials and equipment as well as all of the contractor's tools, construction machinery, and equipment for construction in case that his losses caused by the contractor will not be recovered. The value of the above plus any other sums due to the contractor for work executed in accordance with the contract is determined by the engineer right after the notice of termination has taken effect. The employer may also accept assignment of any of the subcontracts he chooses to accept.

After termination, the employer may complete the Works himself or arrange other entities to do so. The employer and these entities may then use any goods and documents got from the original contractor for the completion of the construction.

d. **payment after termination**

After termination and during the completion of the contract, the employer is not obligated to pay any further funds to the contractor. Instead, he could precede claims against the contractor and withhold further payment until the costs of execution, completion and remedying of any defects, damages for delay in completion, and all other extra costs incurred from the contractor's default have been recovered. In the process of continuing the project, the employer must make sure not to waste funds or incur any upgrading expense and strictly segregate any project betterment costs, as they cannot be charged to the contractor.

If the losses, damages and extra cost to be recovered exceed the unpaid balance of the contract sum, the excess will be charged to the contractor. Or otherwise, the remaining funds must be paid to the contractor, which is quite uncommon in practice however. The architect is required to issue a certificate determining the final payment from or to the contractor.

In many cases, the employer finds it nearly impossible to collect from a bankrupt or impecunious contractor for damage or cost recovery. Under this situation, the employer is entitled to sell the items collected from the contractor in order to recover this payment. Any balance of the proceeds shall then be paid to the contractor.

12.3.3 Termination for Convenience

Terminating a contract for the employer's convenience is unfair to the contractor, yet such clauses often appear in construction contract, allowing an employer to terminate the contractor's or architect's contract for convenience and without any cause. The termination for convenience provisions provide employer with the ability to rid himself of contractor or architect with whom he no longer desires to work. This is consistent with how the construction industry actually operates. The employer can terminate the work of contractor's or architect's for no reason, yet we will only explore the issue of terminating contractor's work for convenience in this chapter.

Although some countries empower employer the right to terminate for convenience, his ability to exercise the right is actually restrained by the resulting consequences of the actual termination. Because most termination for convenience clauses allow the contractor to recover all the costs for the work he has done up to the time of termination, plus reasonable overhead and, in some countries, profit on Works not executed, which could also be a substantial amount. And the right is also restrained by the purpose of termination, e.g. under Sub-Clause **15.5** of FIDIC Conditions of Contract for Construction, the employer shall not terminate the contract in order to execute the Works himself or arrange another contractor to do so, or just to avoid the termination of the contract from the contractor.

When an employer is entitled to terminate the contract for convenience, the following terms are usually included in the contract:

a. the employer should provide proper notice of termination to the contractor;

b. the contractor should then cease operations as directed in the notice;

c. the contractor is entitled to receive payment for Works he has executed, and costs incurred by reason of such termination, along with reasonable overhead and profit on the Works not executed (in some countries, profit on the Work not executed is not an item that should be included in the indemnity amount).

The following is one example of sub-clause for termination for convenience.

> 19.1 Upon written notice to Contractor, Employer may, without cause, terminate all or any part of this contract. Upon receipt of the notice, Contractor should immediately stop work and stop placing of the orders for materials, equipment and supplies in connection with the performance of the terminated Work.
>
> 19.2 Contractor shall make all reasonable efforts to secure cancellation of existing orders and subcontracts upon terms and conditions that are satisfactory to Employers.
>
> 19.3 After termination, the Contractor shall perform only such work as may be necessary to preserve and protect Work already in progress and shall continue to complete any Works not terminated. Contractor shall protect all materials and equipment on the site, or in transit to the site, that are associated with the terminated work.
>
> 19.4 Should Employer elect to terminate the Contract for convenience under this article, the settlement of all claims of Contractor shall be made as follows:
>
> 19.4.1 Employer shall reimburse Contractor for all reasonable costs incurred after the date of termination for protecting Employer's property.
>
> 19.4.2 Employer shall reimburse Contractor for the proportion of his Contract Price for Work actually completed and accepted by Employer. In no event shall Contractor's total reimbursement exceed the Contract Price.

Words, Phrases and Expressions

commence *vi.* 开始 (start, begin)
contravened *vi.* 违反 (go against, violate)
rehabilitate *vi.* 重建，恢复 (to resume, to recover)
deficiency *n.* 缺陷，短缺 (shortage, insufficiency)
duly *adv.* 适当地，适时地 (properly, moderately)
take effect 生效 (become effective)
be obligated to 有义务 (be bound to)
segregate *vi.* 分离 (depart, separate)
betterment *n.* 改善 (improvement)
balance *n.* 余额 (remaining sum; surplus)
impecunious *adj.* 没有钱的；贫穷的 (poor; needy; impoverished)
proceeds *n.* [pl.] 收入；货款收入，卖得金额；收益 (income, earnings)
restrain *vi.* 克制，抑制，约束 (muffle, restrict)

overhead *n.* 管理费用，开销（expenses，expenditure）

疑 难 词 句

1. In construction contract，prior written notice provisions should be specified and the obligations of both parties on termination set out clearly，i. e. the employer's obligation to pay a contractor and the contractor'sobligations to clear the site，etc.

译文：施工合同中，事先写好的通知书条款应具体，双方就终止承担的义务也应清晰，即雇主有义务对承包商付款而承包商有义务清理现场等等。

2. The value of the above plus any other sums due to the contractor for work executed in accordance with the contract is determined by the engineer right after the notice of termination has taken effect.

译文：以上物品的价值，加上依合同就已完成工作而对承包商的应付款项，由工程师于终止通知生效后立即确定。

3. Instead，he could precede claims against the contractor and withhold further payment until the costs of execution，completion and remedying of any defects，damages for delay in completion，and all other extra costs incurred from the contractor's default have been recovered.

译文：相反，他（雇主）可事先向承包商索赔并不做进一步付款，除非执行、完工及修复缺陷的成本，拖延完工造成的损失，由于承包商的违约造成的所有其他成本得到补偿。

4. The termination for convenience provisions provide employer with the ability to rid himself of contractor or architect with whom he no longer desires to work. This is consistent with how the construction industry actually operates.

译文：随意终止合同条款令雇主可以摆脱他想停止合作关系的承包商或设计师，这与施工行业的实际操作相符。

5. And the right is also restrained by the purpose of termination，e. g. under clause 15. 5 of FIDIC *Conditions of Contract for Construction*，the employer shall not terminate the contract in order to execute the Works himself or arrange another contractor to do so，or just to avoid the termination of the contract from the contractor.

译文：其权利还会被终止的目的所限制，比如，FIDIC《施工合同条件》第 15. 5 款规定，如果雇主的终止是为了自己实施工程或为了安排由其他承包商实施该工程，或仅仅是为了避免承包商对合同的终止，则他将无权根据本款终止合同。

6. The contractor is entitled to receive payment for Works he has executed，and costs incurred by reason of such termination，along with reasonable overhead and profit on the Works not executed（in some countries，profit on the Work not executed is not an item that should be included in the indemnity amount）.

译文：承包商有权对其所完成的项目部分、随意终止合同带来的成本、合理的管理费用以及项目未完工部分本应带来的利润要求付款（在有些国家，项目未完工部分本应带来的利润不包括在赔偿金额中)。

12.4 Termination by Contractor

The contractor has an equal right with the employer to terminate a contract for default, which happens in the situation where the misconduct of employer or of sub-contractors is so serious that the law gives the contractor the right to bring the contract to an end. But since the termination decision is frequently followed by disastrous financial consequences and lengthy litigation, the contractor is more cautious to exercise his right of termination. This chapter only put concern on termination for default of the employer's.

The contractor can elect to terminate a contract with the employer when the latter:

a. fails to pay the contractor an amount due within specified period of time after the expiry of the payment time stated in the contract;

b. becomes bankrupt or going into liquidation; or has a receiving or administration order made against him, compounds with his creditors, or carries on business under a receiver, trustee or manager for the benefit of his creditors, or if any act is done or event occurs which (under applicable Laws) has a similar effect to any of these acts or events;

c. gives notice to the contractor that, for unforeseen economic reasons, he cannot continue to meet his contractual obligations;

d. fails to provide reasonable evidence within certain days after receiving contractor's notice of work suspension in respect of a failure to comply with the agreed financial arrangements.

e. substantially fails to perform his obligations under the contract.

The contractor can terminate the contract in, but not limited to, the above events. If you are careful enough, you may find that the events of a, b, c and d are related to the financial condition and/or obligations of the employer.

There are other circumstances that may incur a termination by the contractor, for instance, if services are suspended, the contractor will want the right to terminate after a fixed period, say 6 months, rather than letting it run on indefinitely.

If any of the above events happens, the contractor may give a certain days' notice to terminate the contract. In the circumstances of prolonged suspension and the employer's bankruptcy, however, the contractor can terminate the contract immediately.

By the end of the specified time limit agreed upon in the contract, the termination begins to take effect, the contractor shall then promptly:

cease all further work, except for such work as may have been instructed by the employer for the protection of life or property or for the safety of the Works.

hand over contractor's Documents, Plant, Materials and other work, for which he has received payment.

remove all other Goods from the Site, except those that are essential for safety, and leave the Site.

Correspondently, the employer should return the Performance Security to the contractor if any and make payment to the contractor, which includes the amount of any loss of profit or other loss or damage sustained by the contractor as a result of this termination and other items otherwise regulated in the contract.

Termination should be very much a last resort, a step that should be taken when there is really no alternatives and, as with suspension of the works, a step that should be taken only after seeking legal advice.

Words, Phrases and Expressions

elect *vt.* 选择做某事，决定（choose to do something, decide）
expiry *n.* 终止；完满；满期（running out, finishing）
prolong *vi.* 延长；拉长，拖长（extend, put off）
correspondently *adv.* 相应地
sustain *vi.* 遭受；忍受（suffer, incur）
resort *n.* 倚靠，凭借；手段（option, alternative, choice）

疑 难 词 句

1. (The Employer) fails to provide reasonable evidence within certain days after receiving Contractor's notice of work suspension in respect of a failure to comply with the agreed financial arrangements.

译文：雇主收到承包商因雇主未能履行约定的财务安排而发出的暂停工作通知书后一定时间内未能提供合理的证据。

2. In the circumstances of prolonged suspension and the Employer's bankruptcy, however, the Contractor can terminate the contract immediately.

译文：然而，面对延长的暂停以及雇主破产的情况时，承包商可以立即终止合同。

3. (The Contractor shall) cease all further work, except for such work as may have been instructed by the Employer for the protection of life or property or for the safety of the Works.

译文：承包商应停止一切后续工作，除非该工作由雇主指示进行以保护生命财产安全或保护项目的安全。

4. Correspondently, the Employer should return the Performance Security to the Contractor if any and make payment to the Contractor, which includes the amount of any loss of profit or other loss or damage sustained by the Contractor as a result of this termination and other items otherwise regulated in the contract.

译文：相应地，雇主应退还履约保证金（如果存在的话）给承包商，并向其付款，款项包括：利润损失、承包商因合同终止而遭受的其他损失或损坏，以及合同规定的其他项目。

5. Termination should be very much a last resort, a step that should be taken when there is really no alternatives and, as with suspension of the works, a step that should be taken only after seeking legal advice.

译文：合同终止应为最后的解决手段，为别无他法时才选择的一步，和项目的暂停一样，应进行法律咨询之后方可采取。

Chapter 13 Alternative Dispute Resolutions

Dispute Resolution (hereinafter referred to as DR) is a highly effective way to manage conflict and resolve disputes between parties. It covers a broad spectrum of resolution options such as negotiation, mediation, arbitration, mini-trial and litigation, with the first three options falling into the scope of non-litigious resolution. Rather than litigation, non-litigious resolution now becomes apparently prominent. Businesses, lawyers, government and other groups are trying different approaches or processes apart from litigation to achieve resolution at various stages.

Generally speaking, this flourishing practice of non-litigious resolution is partly a response to dissatisfaction with the current court system's ability to provide timely, effective and efficient resolution of all disputes. In comparison, non-litigious DR provides alternatives to formal judicial determination by the Courts following traditional, public, adversarial litigation. It seeks to provide parties to a dispute with a mechanism for resolution which is chosen to reflect the particular nature of the dispute and the needs and interests of the parties. But instead of non-litigious DR, we run across the term of Alternative Dispute Resolution more frequently in books, which is virtually a term referring to non-litigious methods of resolving legal disputes. Therefore, in the later part of this chapter, we choose to use Alternative Dispute Resolution (ADR). According to Red Book of FIDIC, where notice of dissatisfaction has been given, both parties shall attempt to settle the dispute amicably before the commencement of arbitration, and by amicable settlement, we usually mean ADR.

13.1 Definition of ADR

An Alternative Dispute Resolution process includes any process or procedure, other than an adjudication by a presiding judge, in which a neutral third party participates to assist in the resolution of issues in controversy, through processes such as early neutral evaluation, mediation, mini-trial, and arbitration.

The driving philosophy of ADR is that people in dispute are best equipped to develop their own solutions. This is made easier when assisted by a trained neutral third party. In New Zealand negotiation and mediation are the most common ADR processes.

ADR originated in the USA in a drive to find alternatives to the traditional legal system, felt to be adversarial, costly, unpredictable, rigid, over professionalized, damaging to relationships, and limited to narrow rights-based remedies as opposed to creative problem-solving.

13.2 Features of ADR

There are different forms of ADR, each carrying certain features. However, it is possible to identify certain features that are peculiar to the entire process of ADR. The common denominator of all ADR methods is that they are faster, less formalistic, cheaper and often less adversarial than a court trial.

The first feature of ADR is that, unlike litigation, it deals with all disputes confidentially, which means that neither the process of dispute resolution nor its outcomes should be revealed to public unless required to by a special law.

The second feature of ADR is its flexibility. ADR is more flexible than usual litigation. Unlike the routine process and the result of further conflicts rising out of litigation, ADR takes people away from habitual, reactive and defensive responses, which ultimately escalate conflict. Instead, they move to proactive and constructive approaches, where problems and conflict can be dealt with effectively and early.

Generally, the process of ADR is not so expensive in comparison with litigation, and it is shorter in time as court proceedings are much more complex. ADR grants a feeling of personal satisfaction to the parties, as they can take a more active part in the process and decide themselves which method of ADR to use. ADR helps to reach the compromise between the parties, while litigation results only in satisfaction of one party. And the last characteristic feature is that, by means of ADR, it is possible to get to the root of a problem due to the use of personal approach. In ADR, the decision-making is the responsibility of the disputing parties or the neutral.

ADR has gained widespread acceptance among both the general public and the legal profession in recent years. In fact, before permitting the parties' cases to be tried, some courts even require them to resolve the conflict by ways of ADR. In England, after the April 1999 Lord Woolf's reform to the Civil Procedure, judges have been given extensive power of case management, including: ***"encouraging the parties to use ADR procedure if the Court considers that appropriate, and facilitating the use of such procedure".***

Though ADR has many obvious advantages as compared to usual litigation, there are some cases where it cannot be used, for example, it is not applicable where a definite and broadly applicable solution is required.

Words, Phrases and Expressions

spectrum *n.* 一系列，幅度，范围（continuous range）

mediation *n.* 调解（intermediation）
arbitration *n.* 仲裁（arbitrage）
mini-trial *n.* 微型审判程序
conciliation *n.* 调解，抚慰（mediation）
amicably *adv.* 友好地，和蔼地（agreeably，friendly）
controversy *n.* 争议（dispute，argument）
denominator *n.* 共同特性（common characteristics）
confidentially *adv.* 秘密地（in secret，behind the scenes）
escalate *vi.* 使逐步上升，逐步扩大（to grow rapidly，to heighten）
proactive *adj.* 主动的（initiative）
civil procedure 民事诉讼
dispute resolution 争端解决
Alternative Dispute Resolution 替代性争端解决方法
non-litigious resolution 非诉讼争端解决
presiding judge 审判长；首席法官
over professionalized 过度专业化的

疑 难 词 句

1. It seeks to provide parties to a dispute with a mechanism for resolution which is chosen to reflect the particular nature of the dispute and the needs and interests of the parties.

译文：它（替代性争端解决办法）寻求为争议双方提供一种争端解决机制以反映争端的特殊性质以及争议双方的需要及各自利益所在。

2. An Alternative Dispute Resolution process includes any process or procedure，other than an adjudication by a presiding judge，in which a neutral third party participates to assist in the resolution of issues in controversy，through processes such as early neutral evaluation，mediation，mini-trial，and arbitration.

译文：非诉讼争端解决机制是一种不同于由审判长判决的程序，在该机制下，中立的第三方参与进来，通过早期公正的评估、调节、微型审判程序、仲裁等程序协助解决争端问题。

3. ADR originated in the USA in a drive to find alternatives to the traditional legal system，felt to be adversarial，costly，unpredictable，rigid，over professionalized，damaging to relationships，and limited to narrow rights-based remedies as opposed to creative problem-solving.

译文：替代性争端解决办法是在人们努力寻找新方法以替代传统法律制度的过程中起源于美国的，传统法律制度被认为对抗性强、成本高昂、难以预计、过于刚性化、专业化程度过高、破坏双方关系以及受限于狭隘的权利为基础的补救办法，而不是具有创造性的问题解决机制。

4. In England, after the April 1999 Lord Woolf's reform to the Civil Procedure, judges have been given extensive powers of case management, including: "*encouraging the parties to use ADR procedure if the Court considers that appropriate, and facilitating the use of such procedure*".

译文：在英国，自从1999年4月的沃尔夫民事诉讼司法改革之后，法官被赋予了宽泛的案例管理的权利，包括他可以"如果法庭认为合适，鼓励争议双方使用替代性争端解决办法，并推动该方法的应用"。

13.3 Forms of ADR

ADR is generally classified into at least four types: negotiation, mediation, conciliation and arbitration. Theoretically, arbitration is a form of ADR, but in some countries, it is rather difficult to discern any significant differences between arbitration and litigation. So we will discuss arbitration independently in next chapter.

13.3.1 Negotiation

Negotiation can be defined as any form of direct or indirect communication whereby parties who have opposing interests attempt to reach a compromise by discussing the form of any joint action which they might take to manage and ultimately resolve the dispute between them. A negotiator should advance the interests of the party that he or she represents in order to obtain an optimal outcome for that party.

Traditionally, negotiation occurs directly between the parties and their counsel and does not involve a neutral third party. If the negotiations break down and/or reach an impasse, a third party may be introduced creating a process of facilitated negotiation. Facilitated negotiation tends to be a more ad hoc and informal process than mediation.

Negotiation requires parties to bargain, exchanging information in an attempt to reach a solution. The method provides the advantages of both flexibility and informality. The parties can begin bargaining discussions at the outset of the dispute. Settlement discussions are controlled entirely by the parties and the dispute is resolved only by a solution satisfactory to all.

13.3.2 Conciliation

Conciliation is one form of ADR whereby the parties to a dispute agree to utilize the services of a conciliator, an impartial third party, who then meets with the parties separately and try to resolve their differences. But if a party rejects an offer to conciliate, there can be no conciliation. Conciliation differs from arbitration in that the former has no legal standing, and the conciliator usually has no authority to seek evidence or call witnesses, neither can they make any award. But it assists the parties by driving their negotiations and directing them towards a satisfactory agreement.

When dispute arises and where conciliation is agreed upon previously, parties may submit statement to the chosen conciliators describing the general nature of the dispute and the points at issue. One conciliator is preferred but two or three are also allowed. In case of multiple conciliators, all must act jointly. Each party should also send a copy of the statement to the other. In the process of conciliation, the conciliator may request further details, ask to meet the parties, or communicate with them orally or in writing. Because of trust, parties may even submit to the conciliator suggestions for the settlement of the dispute.

When it appears to the conciliators that elements of settlement exist, they may draw up terms of settlement and send them to the parties for their acceptance. If both the parties sign the settlement document, it shall be final and binding on both.

In USA, this process is similar to mediation. However, in India, conciliation is different from mediation and is a completely informal type of ADR mechanism.

13.3.3 Mediation

While each of the ADR processes may be effective in various circumstances, mediation is now widely recognized in the UK, Europe and the United States as the most popular form of ADR, because it offers superior advantages for the resolution of disputes and affords the parties the opportunity to develop settlements that are practical, economical, and durable. The disadvantage is that there is not guarantee of resolution. (However, effective mediations result in resolutions around 95% of time.)

In many countries, mediation is a concept that sounds strikingly similar to conciliation, so it is often mistakenly confused with conciliation. In Europe and some areas of the United States, what is termed as "mediation" is known as "conciliation". Although the two methods have similar aspects, they are fundamentally different ADR approaches.

Like conciliation, the mediation process also involves an independent third party helping disputing parties to resolve their dispute. The disputants, not the mediator, decide the terms of the agreement, while a mediator assists the parties in identifying and articulating their own interests, priorities, needs and wishes to each other by communicating with each party. A purely facilitative mediator expresses no opinion on the matter, while an evaluative mediator will express opinions and perhaps advise the parties on ways to resolve the case.

Conciliation and mediation are similar in that they both look to maintaining an existing business relationship rather than ruining it like in litigation and they both have impartial third party helping in the process. And both the conciliator and the mediator may have an advisory role in regard to the content of the dispute or the outcome of its resolution, but not a determinative role, which is distinct from the role of arbitrators'.

So lots of times we can see these two concepts used as synonyms. But the truth is they do vary substantially in both their procedures and the role played by the conciliator and mediator.

a. the difference in procedure

In mediation, there are several stages to follow, i. e. , introduction, joint session, caucus, and agreement, and the mediator controls the process through different and specific stages but not the outcome, which is left to the parties. By contrast, in conciliation, there may not be a structured process to follow. The conciliator may instead administer the conciliation process as a traditional negotiation, which may take different forms depending on the case.

b. the difference in the role of conciliator and mediator

The conciliator and the mediator also play different role in ADR. Firstly, While the conciliator, as the impartial party, is responsible for figuring out the best solution, the mediator is not. Secondly, The conciliator, not the parties, often develops and proposes the terms of settlement; by contrast, the mediator helps to facilitate the parties' own discussion and guide them to their own solution. Thirdly, the conciliator may make proposals for settlement, while in mediation, the decisions are made by the conflicting parties. To sum up, in conciliation, the conciliator plays a relatively direct role in the actual resolution of a dispute and is usually seen as an authority, but in mediation, the mediators at all times maintain their neutrality and are considered as a partner to give assistance.

c. the difference in the timing

Conciliation is used almost before a conflict grows to serious one, and its aim is to stop a substantial conflict from developing. But mediation happens when the conflict has already surfaced and grown into a serious one that is difficult to deal with without "professional" assistance.

13. 3. 4 Dispute Adjudication Board

The approach of Dispute Board is a highly successful and cost effective contractual process for the avoidance and resolution of disputes during the course of the project. The two most popular DB types are: the Dispute Review Board (DRB) which originated in the USA and aims at providing non-binding recommendations and the Dispute Adjudication Board (DAB) which emerged from the earlier USA mode and is entitled to make decision that has an interim binding force. There is a new emerging type of DB namely Combined Dispute Board (CDB), which is a hybrid of Dispute Review Boards and Dispute Adjudication Boards created by the International Chamber of Commerce (ICC) in 2004. In practice, there are also other terms in use such as Dispute Settlement Panel, Dispute Mediation Board, Dispute Avoidance Panel and Dispute Conciliation

Panel. Fundamentally these different varieties of Dispute Board devices are the same, each providing early adjudication based on the contractual bargain between the parties.

A DRB or DAB is a "job-site" dispute adjudication process, typically comprising three independent, experienced, impartial and respected construction professionals selected by the contracting parties (For smaller projects, the Board may comprise just one single member). The board, in different cases, could be a standing or an *ad hoc* one. The significant differences between DRB/DAB and most other ADR techniques are as the following:

- The Board is appointed at the commencement of a project before any disputes arise.
- The professionals undertake the responsibility of regularly visiting to the site and are actively involved throughout the project (and possibly any agreed period thereafter). So, DRB/DAB is, in some way, a part of the project administration.
- As a part of the project administration, this approach can influence, during the contract period, the performance of the contracting parties.
- DRB/DAB offers expedite and potentially cost effective dispute resolution.
- Despite the interim binding decision of DAB, there will still be the opportunity of referring to arbitration or the courts if the Board's decision is unacceptable to the parties.

This approach is ideally suited to larger international projects, multi-contract projects and concessions. It had been adopted by influential employers such as AT&T and Hawaiian Department of Transportation and applied in projects like El Cajon Dam and Hydropower Project in Honduras in 1980. The *Master Bidding Document for Procurements of Works and User's Guide* (July 2005) adopted by World Bank regulates the use of DB. Following the World Bank example in 1995/96, International Federation of Consulting Engineers (FIDIC) introduced "Dispute Arbitration Board" in its Orange Book (Contract for Design-Build and Turnkey) and optional amendment to standard form contracts. FIDIC made further revision in 1999 making DB the principal means of dispute resolution (the Red Book involves the establishment of a standing DB and the Yellow Book and Silver book involve establishment of the DAB deferred until disputes arise). Further, the Gold Book decided the form of standing DB in 2008.

Considering all the advantages, many international institutions (eg. World Bank, FIDIC and ICC) advocate the wide use of DB, and the use of DRB/DAB continues to grow in worldwide popularity. In UK, influential projects using Dispute Boards are Channel Tunnel Rail Link, London Docklands Light Railway and various Highways Agency projects. The latest conducted research indicates that almost 98% of disputes referred to Dispute Boards conclude the matter without recourse to arbitration or litigation.

Words, Phrases and Expressions

counsel *n.* 律师，辩护人（lawyer，defender）
impasse *n.* 僵局，陌路（deadlock，dead end）
facilitated negotiation *n.* 协助谈判
ad hoc adj. 特定，专设，临时（temporary）
reject *vi.* 拒绝（to turn down，to refuse）
legal standing *n.* 法律地位（legal status）
disputants *n.* 争议者（wrangler）
articulate *vt.* 阐明（to clarify，to explain clearly）
facilitative mediator 促进式调解人
evaluative mediator 评估式调解人
synonyms *n.* 别名（another name，byname）
joint session【法】联席会议（joint meeting，joint conference）
caucus *n.* 预备会议；核心小组（ginger group）
Dispute Board 争端委员会
Dispute Adjudication Board 争端裁定委员会
Dispute Review Board 争议评审委员会
interim *adj.* 暂时的，临时的（provisional，temporary）
hybrid *n.* 混合（mix，interfusion）
job-site *n.* 施工现场（work site，construction site）
concession *n.* 特许项目（license，special permission）
standing board 常设争议小组
Dispute Settlement Panel 争端解决小组
Dispute Mediation Board 争议调解委员会
Dispute Avoidance Panel 争端避免小组
Dispute Conciliation Panel 争端调解小组
The Master Bidding Document for Procurements of Works and User's Guide 世界银行采用的《工程采购和用户指南主导招标文件》

疑 难 词 句

1. Negotiation can be defined as any form of direct or indirect communication whereby parties who have opposing interests attempt to reach a compromise by discussing the form of any joint action which they might take to manage and ultimately resolve the dispute between them.

译文：谈判可以定义成任何直接或间接的沟通形式，在该形式下，有着不同利益的争议双方试图通过讨论他们可能采取的联合行动的方式达成和解以求最终解决争议。

2. When dispute arises and where conciliation is agreed upon previously，parties may

submit statement to the chosen conciliators describing the general nature of the dispute and the points at issue.

译文：当争端产生且双方事先同意使用和解方法时，争议双方可向选定的和解人递交陈述函以描述争端的大致性质和双方关注焦点。

3. because it offers superior advantages for the resolution of disputes and affords the parties the opportunity to develop settlements that are practical, economical, and durable.

译文：因为它为争议解决提供了更先进的方法，并且为争议双方提供了探索实用的、经济的和持久的解决办法的机会。

4. The disputants, not the mediator, decide the terms of the agreement, while a mediator assists the parties in identifying and articulating their own interests, priorities, needs and wishes to each other by communicating with each party.

译文：合同条款由争议人而非调解人制定，调解人通过与双方沟通，协助争议双方向对方确认并阐明自己的利益、需要优先解决的问题、各自的需要及愿望等。

5. And both the conciliator and the mediator may have an advisory role in regard to the content of the dispute or the outcome of its resolution, but not a determinative role, which is distinct from the role of arbitrators'.

译文：和解人与调解人都能在争议内容或解决结果方面起到咨询作用而不是决定性作用，这点区别于仲裁人的作用。

6. The two most popular DB types are: the Dispute Review Board (DRB) which originated in the USA and aims at providing non-binding recommendations and the Dispute Adjudication Board (DAB) which emerged from the earlier USA mode and is entitled to make decision that has an interim binding force.

译文：两种最典型的争端委员会是：一为争端评审委员会，该委员会起源于美国，致力于提供不具约束力的建议，二为争端裁定委员会，该委员会从美国早期的模式中发展而来，并有权作出具有临时约束力的裁定。

7. Despite the interim binding decision of DAB, there will still be the opportunity of referring to arbitration or the courts if the Board's decision is unacceptable to the parties.

译文：尽管争端裁定委员会的裁定具有临时约束力，但是如果委员会的裁定无法为争端双方所接受，他们仍有机会将其提交仲裁或法庭。

13.4 The Mediation/ Conciliation Agreement

In spite of the differences between mediation and conciliation, the two approaches are similar in the content and the form of their agreement. The mediation agreement needs to be clear so that both parties know what to expect from mediation. Usually, the agreement should include at least the following matters:

• A clear and precise statement of what kind of issues should be solved through mediation/ conciliation.

• A declaration by the parties that they both wish to use mediation/ conciliation to solve their dispute before it is referred to arbitration or the court.

• A period during which a mediation/ conciliation is to take place.

• The name and the qualification of the mediator/conciliator.

• A date for a meeting with the mediator/conciliator to discuss format.

• How cost arising out of mediation/ conciliation is to be shared between the parties.

• The statement that the mediation/ conciliation is to be conducted on a confidential basis, and a decision that the mediator/conciliator will not be called to give evidence in any subsequent proceedings.

13.5 The Procedure of DRB/DAB

13.5.1 Selection and Appointment of Dispute Board Members

The members in a Dispute Board should be impartial and objective, they should be skilled and experienced in this area and be able to work in harmony with other members. When selecting the members, the parties should be prudent to use an odd number. In most cases, there are 3 members, but in other cases, the number could also be 1 or over 3 depending on the size of a project.

Typically, each party can select one member of the DB with the third (usually seen as chairman) either jointly appointed by the parties or by the other two selected members. Each party holds the right to lodge reasonable objection to the other party's selection. And the number could be either appointed by the parties themselves or suggested by influential institutions such as ICC. The members could be resigned or replaced in the latter process of "administration", but only upon the term of DB agreement.

13.5.2 Dispute Board Rules

The rules could be decided by the parties or the members of the Board under permission of the parties. The parties may also select standard forms and institutional rules made by institutions like ICC. According to the rules, the cost is usually shared between the contracting parties.

13.5.3 The Work of the Dispute Board

The Dispute Board members, once appointed, are kept informed of construction progress and development, and are allowed to visit the site periodically (typically every

3 months) to be familiar with the specifics of the project and of the parties. During the visits they observe any problems as (and before) they develop, and discuss matters of concern with the parties.

Should any dispute arises, the decision should be firstly made by the engineer and the referral to DB occurs only upon dissatisfaction of any parties. The Dispute Board can informally assist the parties in trying to resolve any disagreements before they escalate into adversarial disputes.

Once referred to a DB by any party, there should be a following stage of hearing, in which the members consider the issues and provide a written recommendation or decision. If the parties resort to quick solution of the dispute, it is important to specify a time limit within which the DB is to make its decision. Example time limit in standard forms range from 14 days (AAA), 84 days (World Bank and FIDIC) to 90 days (ICC).

By the end of time limit, the board should give recommendations, interim binding or non-binding. Interim binding decisions are those that might be binding but pending final determination of the dispute before referred to a court or an arbitral tribunal. Non-binding recommendations are recommendations with which the parties are free to comply or reject and proceed directly to arbitration or litigation. And the decision is not capable of enforcement under the New York Convention nor does it have the status of a court judgement.

In recent years the term Alternative Dispute Resolution has begun to lose favor in some circles and been replaced by the term Appropriate Dispute Resolution. Because even the judiciary suggests the use of non-litigious resolution before the submission of disputes to the court. In reality, fewer than 5 percent of all lawsuits filed go to trial; the other 95 percent are settled or otherwise concluded before trial. As Alternative Dispute Resolution processes make headway into our legal system and business practice, it seems inaccurate to call them "alternative" any more. The point of this semantic change is to emphasize that ADR methods stand on their own as effective ways to resolve disputes and should not be seen simply as alternatives to a court action.

Words, Phrases and Expressions

odd *adj.* 奇［单］数的（uneven, (of a number) that can not be exactly divided by 2)

lodge *vi.* 正式提出（to make a statement officially, to offer formally)

hearing *n.* 听证会（a chance to be heard explaining one's position)

tribunal *n.* 法庭（court)

semantic *adj.* 语义的（of or related to meaning language)

Appropriate Dispute Resolution 合适的争端解决办法

court action 法律手段

疑 难 词 句

1. Typically, each party can select one member of the DB with the third (usually seen as chairman) either jointly appointed by the parties or by the other two selected members.

译文：典型的做法是：每一方各选定一名小组成员，第三名小组成员（首席）由双方当事人共同选定或由当事人选定的小组成员共同选定。

2. The Dispute Board members, once appointed, are kept informed of construction progress and development, and are allowed to visit the site periodically (typically every 3 months) to be familiar with the specifics of the project and of the parties.

译文：一旦指定了争议委员会成员，就应该时刻向其告知工程进展，并允许其定期（通常是每 3 个月一次）巡视施工现场，以便熟悉项目以及双方细节。

3. Should any dispute arises, the decision should be firstly made by the engineer and the referral to DB occurs only upon dissatisfaction of any parties.

译文：一旦发生争议，应首先由工程师进行裁定，只有任一方对裁定不满时才可将争议提交争议委员会。

4. Interim binding decisions are those that might be binding but pending final determination of the dispute before referred to a court or an arbitral tribunal. Non-binding recommendations are recommendations with which the parties are free to comply or reject and proceed directly to arbitration or litigation.

译文：临时约束性建议在提交法院或仲裁法庭之前其约束性有待确定；而对于争议小组做出的非约束性建议，当事人可以选择遵从或拒绝，并直接提起仲裁或诉讼。

5. And the decision is not capable of enforcement under the New York Convention nor does it have the status of a court judgement.

译文：依据纽约公约，争议小组决定既不具有强制执行力，也不具有与法院判决同等的地位。

6. The point of this semantic change is to emphasize that ADR methods stand on their own as effective ways to resolve disputes and should not be seen simply as alternatives to a court action.

译文：这种语义上的变化（由 Alternative Dispute Resolution 到 Appropriate Dispute Resolution）意味着替代性争端解决办法作为一种有效的争议解决方式自成体系，而不应仅被看做是法律手段的替代方式。

Chapter 14 Arbitration and Litigation

14.1 Arbitration

"*For an arbitrator goes by the equity of a case, a judge by the law, and arbitration was invented with the express purpose of securing full power for equity.*" Through his quote, Aristotle has clearly defined the paramount role that arbitration plays in seeking justice. Arbitration is a proceeding in which a dispute is resolved by an impartial adjudicator whose decision the parties to the dispute have agreed will be final and binding.

Arbitration is a very common approach for resolving international commercial or business disputes, and a number of international organizations have been established for doing this. These include the International Chamber of Commerce (in Paris), the Arbitration Institute of the Stockholm Chamber of Commerce, and the International Court of Arbitration administered by the World Business Organization. Each of these bodies has its own procedure for arbitration which, avoids the problem of one country having different procedure from another.

There are something in common between arbitration and mediation, for both have impartial parties involved in the proceeding. But the difference is apparent. The most basic difference is that arbitration involves a decision by the neutral arbitrator after an evidential hearing where the arbitrator is typically a passive participant whose role is to determine right or wrong, whereas the mediator by contrast, is generally an active participant who attempts to lead the parties to reconciliation and agreement, regardless of who is right or wrong. At the end of mediation, the parties write up their own settlement agreement which can be temporarily binding or nonbinding at all with the assistance of the mediator; and at the end of arbitration, the arbitration tribunal will make an award which is usually binding.

14.1.1 Advantages of Arbitration over Litigation

Among the available dispute resolution alternatives to the courts, arbitration is by far the most commonly used internationally. Many of the advantages most frequently claimed of arbitration are especially relevant to those rising from construction contract. The reasons for this are clear:

a. Decisions being final and binding

Arbitral Awards have the feature of enforceability. While several mechanisms can help parties reach an amicable settlement (for example through mediation under the ICC ADR Rules) all of them depend, ultimately, on the goodwill and cooperation of the parties. Once any of the parties is dissatisfied with the resolution, he will still turn to courts or arbitration for a final and enforceable decision. But comparatively, the decision of arbitration is final and binding. Because arbitral awards are not subject to appeal, they are much more likely to be final than the judgments of courts of first instance. Although arbitral awards may be subject to being challenged (usually in the country where the arbitral award is rendered or where enforcement is sought), the truth is awards can be attacked only under very limited circumstances. So unlike court judgments, arbitration awards are enforceable world wide in accordance with some prevailing conventions such as the Convention on the Recognition and Enforcement of Foreign Arbitral Awards, known as the New York Convention.

b. Arbitral Awards Being Internationally Recognized

Arbitral awards have gained much greater international recognition than judgment of national courts. Let' s take for instance the New York Convention which was adopted in 1958 by the United Nations General Assembly. By the end of January, 2009, 144 countries including China have signed the Convention. It facilitates enforcement of awards in all contracting states. According to Article 1 of the Convention, "*This Convention shall apply to the recognition and enforcement of arbitral awards made in the territory of a State other than the State where the recognition and enforcement of such awards are sought, and arising out of differences between persons, whether physical or legal. It shall also apply to arbitral awards not considered as domestic awards in the State where their recognition and enforcement are sought.*" There are several other multilateral and bilateral arbitration conventions that may also help enforcement.

c. Neutrality

The feature of neutrality is reflected in a neutral arbitrator, a neutral place and a neutral proceeding.

In international arbitration, for example, the rule is firm that arbitrators must act in an independent and impartial manner, unless the parties agree otherwise. In some arbitrations, party-appointed arbitrators in the panel are expected by each party to act as "non-neutrals" and may be predisposed to decide in favor of the party who appointed them, while the third arbitrator is neutral.

The choice of the place of arbitration in particular is extremely important. This is because the place of arbitration determines which procedural law will apply to the arbitration. The oral proceedings, on the other hand, may take place anywhere. If, for instance, Amsterdam is chosen as the place of arbitration, the arbitration will be subject to Dutch procedural law. The oral proceedings can then take place elsewhere in the Netherlands or even in another country.

Arbitration proceedings are neutral. In litigation, parties are confronted by a lack of choice and most of the times are forced to use the national courts at hand to settle their disputes. Furthermore, due to the absence of a truly international court for the resolution of transnational commercial disputes, cases are usually instituted in the courts of the state where either the claimant or the defendant resides. On the other hand, arbitration offers the possibility of a completely neutral panel of adjudicators. This, in turn, eliminates the possibility of actual or perceived bias on the part of the court.

Furthermore, arbitration may take place in any country, in any language and with arbitrators of any nationality. With this flexibility, it is generally possible to structure a neutral procedure offering no undue advantage to any party.

d. Freedom in Choice of Arbitrator

Judicial systems do not allow the parties to a dispute to choose their own judges. In contrast, arbitration offers the parties the unique opportunity to designate persons of their choice as arbitrators, provided they are independent. This freedom to select the arbitrator is why arbitration has been described as "hiring your own private judge." This enables the parties to have their disputes resolved by people who have specialized competence in the relevant field. Because a proper understanding of any disputes arising from construction contract can only be gained by long experience in the construction industry, it is desirable that the arbitrator in such cases should be an experienced engineer (or, where appropriate, architect or quantity surveyor).

In international arbitration, there is usually a tripartite panel designated. The system allows each party to choose its own arbitrator on the panel who will make sure the party's position is clearly and effectively presented to the third arbitrator who stands in a totally neutral position. The party-appointed arbitrators mainly help to clarify technical issues and make sure that the neutral arbitrator fully understands the issues and background of the case, the intentions of each party before any award is given. Throughout the course of the arbitration, when the three arbitrators confer privately during executive sessions, they combine various viewpoints on the subject area. This exchange of ideas should lead to a more thorough and comprehensive decision. Also, an award rendered by three persons is often more acceptable to the parties and recourse to the courts is therefore less likely.

e. Speed and Economy

Arbitration is faster and less expensive than litigation in the courts. Overburdened and congested national courts unnecessarily elongate the length of time required to resolve a dispute. Once the parties have entered the litigation process they will find themselves involved in unavoidable time-consuming stages starting from the Court of First Instance, through to the Court of Cassation in order to come to a final judgment. Parties in commercial disputes can not afford to wait; they will need to resolve their disputes in the fastest manner they find suitable. Although a complex international dispute may sometimes take a long time and a great deal of money to resolve, even by arbitration, the limited scope for challenge against arbitral awards, as compared with court judgments, offers a clear advantage. The costs for the arbitration process are limited to the fee of the arbitrator (depending on the size of the claim, expertise of the arbitrator, and expenses), and attorney fees. Costs for litigation include attorney fees and court costs, which can be very high. Above all, arbitration helps to ensure that the parties will not subsequently be entangled in a prolonged and costly series of appeals. Furthermore, arbitration offers the parties the flexibility to set up proceedings that can be conducted as quickly and economically as the circumstances allow. In this way, a multi-million dollar ICC arbitration was once completed in just over two months.

f. Confidentiality

Compared with the publicity of court proceedings and judgments, arbitrations and awards are normally private, and only the parties themselves receive copies of the awards.

Arbitration proceedings, unlike those in the courts, are not open to the pressure or to the public; only those persons involved in the proceedings are entitled to attend. Some other persons may also be allowed to be present in some special cases , for example when an arbitrator wishes a pupil to gain experience. But such attendance is normally on the condition that the confidentiality of the proceedings will be respected.

Usually neither parties in dispute wishes to publicize the matter disclosed at the hearing; and frequently the damage to the previous harmonious relationship between two parties resulting from a dispute is more rapidly healed where there is no publicity.

Words, Phrases and Expressions

paramount *adj.* 重要的，首要的（supreme, chief, vital）
reconciliation *n.* 调停，和解（settlement, compromise, ceasefire）
prevail *vi.* 获胜，有优势（win through, succeed）
appeal *n.* 起诉，上诉（taking a case to a higher court for review）
accede *vi.* 同意加入（enter upon）
convention *n.*（国际）公约，协定（pact, agreement or treaty）
neutrality *n.* 中立（impartiality, refusal to take sides）
claimant *n.* 原告，债权人（plaintiff, accuser）
defendant *n.* 被告（the person who is accused）
reside *vi.* 住，居住（in; at）（be located in, inhabit）

institute *vi.* 开始（调查），提起（诉讼）(lodge a complaint)
panel *n.* 小组（team，small group of people）
tripartite *n.* 三个一组的，三方契约（a team constituted with three persons）
confer *vi.* 商议，协商，谈判（put heads together，negotiate）
congest *vi.* 充满，拥塞（crowed，jam-packed）
elongate *vi.* *vi.* 拉长，（使）延长（lengthen，make longer ）
attorney *n.* ［美国］辩护律师（lawyer）
confidentiality *n.* 机密性，保密性（the state of being secret）
United Nations General Assembly 联合国大会
be predisposed to 本来爱好…，有…的倾向（be inclined to ）
executive session 执行会议
Court of First Instance 高等法院原讼法庭
Court of Cassation 最高法院
the Convention on the Recognition and Enforcement of Foreign Arbitral Awards 联合国承认和执行外国仲裁裁决的公约

疑　难　词　句

1. At the end of mediation, the parties write up their own settlement agreement which can be temporarily binding or nonbinding at all with the assistance of the mediator; and at the end of arbitration, the arbitration tribunal will make an award which is usually binding.

译文：调解结束之际，（冲突）双方在调解人的协助下达成和解协议，该协议可以有临时的约束力或根本没有约束力；而仲裁结束之际，仲裁庭将做出判决，而该项判决通常是有约束力的。

2. This Convention shall apply to the recognition and enforcement of arbitral awards made in the territory of a State other than the State where the recognition and enforcement of such awards are sought, and arising out of differences between persons, whether physical or legal. It shall also apply to arbitral awards not considered as domestic awards in the State where their recognition and enforcement are sought.

译文：由于自然人或法人之间的争执而引起的仲裁裁决，在一个国家的领土内达成后，而在另一个国家请求承认和执行时，适用本公约。在一个国家请求承认和执行这个国家不认为是本国裁决的仲裁裁决时，也适用本公约。

3. In litigation, parties are confronted by a lack of choice and most of the times are forced to use the national courts at hand to settle their disputes. Furthermore, due to the absence of a truly international court for the resolution of transnational commercial disputes, cases are usually instituted in the courts of the state where either the claimant or the defendant resides.

译文：在诉讼中，双方没有更多选择，更多时候只能被迫采用本国法庭来解决纠纷。进一步说，由于缺乏真正意义上的国际法庭来解决跨国商业纠纷，所以他们一般都在原告或被告所在的国家的法院将案件提起诉讼。

4. Although a complex international dispute may sometimes take a long time and a great deal of money to resolve, even by arbitration, the limited scope for challenge against arbitral awards, as compared with court judgments, offers a clear advantage.

译文：虽然通过仲裁来解决一个比较复杂的国际争端有时也会需要一个漫长的时间和大量的金钱，但考虑到仲裁裁定可能受到质疑的局限性，与法院判决相比，仲裁仍具有较明显的优势。

5. Usually neither of the parties in dispute wishes to publicize the matter disclosed at the hearing; and frequently the damage to the previous harmonious relationship between two parties resulting from a dispute is more rapidly healed where there is no publicity.

译文：通常争议的任何一方都不希望公开听证会上的内容；在不公开情况下，双方因纠纷而遭到破坏的以往和谐的关系往往会更快地得到恢复。

14.1.2 Disadvantages of Arbitration

In spite of all the advantages of arbitration, some people still prefer the court procedure, for in practice, arbitration did show some disadvantages. As mentioned above, arbitration has so many advantages comparing with the litigation. But it still has very serious shortcomings.

Firstly, some people argue that the arbitration process may not be fast when there are multiple arbitrators on the panel, juggling their schedules for hearing dates in long cases which can lead to delays. And it may not be inexpensive either, for in some arbitration agreements, the parties are required to pay for the arbitrators, which adds an additional layer of legal cost that can be prohibitive. And in this case, slow processing of arbitration will waste time and lead to higher expenditure.

Secondly, arbitrators' decisions are difficult to overturn because judicial review of an arbitrator's decision is limited. This may create risk, especially when crucial discovery may be barred and relaxed evidentiary rules may lead to undesirable and otherwise inadmissible evidence. In court, there also is the ability to appeal, but the ability to appeal an arbitrator's decision is extremely limited, which means that an erroneous decision cannot be easily overturned.

Thirdly, in some legal systems, arbitral awards have fewer enforcement remedies than judgments; although in the United States, arbitration awards are enforced in the same manner as court judgments and have the same effect.

Fourthly, arbitrators are generally unable to enforce interlocutory measures against a party, which makes it easier for a party to take steps to avoid enforcement of an award, such as the relocating assets offshore.

Fifthly, arbitration is adversarial, thus it generally does nothing to create win-win solutions or improve relationships. Often it escalates a conflict, just as court-based adjudication is likely to do. In addition, arbitration takes decision making power away from the parties. This results in a resolution of the current conflict, but does nothing to help the parties learn how to resolve their own conflicts more effectively in the future, as does mediation.

At present, because of these disadvantages, arbitration has been facing significant competition over the last few years from new tools of Alternative Dispute Resolution (ADR) such as negotiation, mediation, conciliation, and mini-trials. These ADR mechanisms have been consistently advanced as alternative means either to complement arbitration or to replace it altogether. So, arbitration institutions must make use of the advantage of new digital technology to short the processing, decrease the inconvenience and low the expense.

14.1.3 Arbitration Institutions

Arbitration may be commenced prior to or after completion of the Works, provided that the obligations of the employer, the engineer and the contractor shall not be altered by reason of the arbitration being conducted during the progress of the Works.

Parties using arbitration have a choice between designating an institution, such as ICC, to administer it, or proceeding *ad hoc* outside an institutional framework. In ad hoc cases, the arbitration will be administered by the arbitrators themselves. However, should problems arise in setting the arbitration in motion or in constituting the Arbitral Tribunal, the parties may have to require the assistance of a state court, or that of an independent appointing authority such as ICC. Although institutional arbitration requires payment of a fee to the administering institution, the functions performed by the institution can be critical in ensuring that the arbitration proceeds to a final award with a minimum of disruption and without the need for recourse to the local courts.

In general, there are two broad categories of the arbitration institutions, namely international institutions based on international convention such as ICSID (International Centre for Settlement of Investment Disputes) and Commission on Arbitration of ICC and local institutions registered in particular country.

International Institutions

- ***ICSID***

ICSID is an autonomous international institution established under the Convention on the Settlement of Investment Disputes between States and Nationals of Other States (the ICSID or the Washington Convention) with over 140 member States. The ICSID

Convention is a multilateral treaty formulated by the Executive Directors of the International Bank for Reconstruction and Development (the World Bank). The Convention sought to remove major impediments to the free international flows of private investment posed by non-commercial risks and the absence of specialized international methods for investment dispute settlement. ICSID was created by the Convention as an impartial international forum providing facilities for the resolution of legal disputes between eligible parties, through conciliation or arbitration procedures. Recourse to the ICSID facilities is always subject to the parties' consent. Today, ICSID is considered to be the leading international arbitration institution devoted to investor-state dispute settlement.

■ ***ICC Commission on Arbitration***

The Commission on Arbitration aims to create a forum for experts to pool ideas and impact new policy on practical issues relating to international arbitration, the settlement of international business disputes and the legal and procedural aspects of arbitration. The Commission also aims to examine ICC dispute settlement services in view of current development, including new technologies.

Local Arbitration Institutions

■ ***American Arbitration Association (AAA)***

The American Arbitration Association (AAA) is a private enterprise in the business of arbitration and one of several arbitration organizations that administers arbitration proceedings. The AAA also administers mediation and other forms of ADR. The International Centre for Dispute Resolution (ICDR), established in 1996, administers international arbitration proceedings initiated under the institution's rules. Many contracts include an arbitration clause naming the AAA as the organization that will administer an arbitration between the parties. The AAA itself does not arbitrate disputes, but provides administrative support to arbitrations before a single arbitrator or a panel of three arbitrators. The arbitrators are chosen in accordance with the parties' agreement or, if the parties do not agree otherwise, in accordance with the AAA rules. Under its rules, the AAA may appoint an arbitrator in some circumstances, for example, where the parties cannot agree on an arbitrator or a party fails to exercise its right to appoint an arbitrator.

■ ***London Court of International Arbitration (LCIA)***

The London Court of International Arbitration (which now goes by the name of its acronym LCIA) is a London based institution providing the service of international arbitration. It is an international institution which provides a forum for dispute resolution proceedings for all parties, irrespective of their location or system of law. Although arbitration is the Institution's main focus, the LCIA is also active in mediation, a form of ADR.

■ ***Arbitration Institute of the Stockholm Chamber of Commerce* (*SCC*)**

The Arbitration Institute of the Stockholm Chamber of Commerce (SCC Institute) was established already in 1917 and has since then been active in both domestic and international arbitration. While at the outset the SCC Institute mainly administered domestic arbitrations, the international activities have increased significantly over the past twenty-five years. Out of about 160 cases in 2004, roughly half the number is international.

The SCC Institute is an entity within the Stockholm Chamber of Commerce. It is, however, functionally independent and decisions of the Board of the Institute are final and not subject to review by the Chamber.

■ ***China International Economic & Trade Arbitration Commission* (*CIETAC*)**

CIETAC has become one of the major commercial arbitration centers around the world. It was established in 1954 within the China Council for the Promotion of International Trade to settle, by means of arbitration, disputes arising from economic and trade transactions, contractual or non-contractual and particularly disputes between foreign and Chinese firms or companies.

■ ***Hong Kong International Arbitration Center* (*HKIAC*)**

HKIAC was established in 1985 to assist disputing parties to solve their disputes by arbitration and by other means of dispute resolution. It was established by a group of leading business and professional people in Hong Kong to be the focus for Asia of dispute resolution. It has been generously funded by the business community and by the Hong Kong Government but it is totally independent of both and it is financially self sufficient.

Words, Phrases and Expressions

overturn *vi. vi.* 颠覆，毁灭，推翻 (overthrow)
bar *vi.* 阻挡，阻挠，防止，排斥 (get in the way, hinder)
inadmissible *adj.* 不能允许的，难承认的 (unacceptable, not permitted)
erroneous *adj.* 错误的，不正确的 (incorrect, wrong, invalid)
interlocutory *adj.* (判决等) 在诉讼期间宣告的，非最后的 (not final)
autonomous *adj.* 自治的，自主的 (self-directed, self-governing)
impediment *n.* 妨碍，阻碍，障碍物 (obstacle, obstruction, hindrance)
acronym *n.* 首字母缩略词
pool *vi.* 共享，集中 (智慧等) (share, to concentrate idea together)
judicial review 司法审查
set something in motion 启动，开始，付诸行动 (start, commence)
Convention on the Settlement of Investment Disputes between States and Nationals of Other States 关于解决各国和其他国家的国民之间的投资争端的公约

ICSID (International Centre for Settlement of Investment Disputes) 国际投资争端解决中心

International Bank for Reconstruction and Development 国际复兴与开发银行

ICC Commission on Arbitration 国际仲裁委员会

American Arbitration Association (AAA) 美国仲裁协会

International Centre for Dispute Resolution (ICDR) 国际争议解决中心

London Court of International Arbitration (LCIA) 伦敦国际仲裁院

Arbitration Institute of the Stockholm Chamber of Commerce (SCC) 斯德哥尔摩商会仲裁院

China International Economic & Trade Arbitration Commission (CIETAC) 中国国际经济贸易仲裁委员会

China Council for the Promotion of International Trade 中国国际贸易委员会

Hong Kong International Arbitration Center (HKIAC) 香港国际仲裁中心.

arbitral tribunal 仲裁法庭

疑 难 词 句

1. Thirdly, in some legal systems, arbitral awards have fewer enforcement remedies than judgments; although in the United States, arbitration awards are enforced in the same manner as court judgments and have the same effect.

译文：第三，在一些法律制度下，仲裁裁决在强制执行的补救措施方面不如判决；尽管在美国，仲裁裁决和法庭判决一样被强制执行，且具有一样的效果。

2. Arbitration may be commenced prior to or after completion of the Works, provided that the obligations of the employer, the engineer and the contractor shall not be altered by reason of the arbitration being conducted during the progress of the Works.

译文：假如雇主、工程师和承包商的义务不会因项目进行过程中进行仲裁而有所改变的话，仲裁可以开始于项目完工之前或之后。

3. However, should problems arise in setting the arbitration in motion or in constituting the Arbitral Tribunal, the parties may have to require the assistance of a state court, or that of an independent appointing authority such as ICC.

译文：然而，一旦在仲裁的启动和仲裁法庭的组成上出现问题，（争议）双方将不得不要求州法院或像国际商会那样的独立的指定机构的协助。

14.1.4 Choosing of Arbitrator

If decision of the Engineer, if any, has not become final and binding, and notice of dissatisfaction has been given to the Dispute Board, both parties shall attempt to settle the dispute amicably before the commencement of arbitration. However, Sub-Clause 20.5 of FIDIC Conditions of Contract for EPC/Turnkey (Silver Book) regulates that: unless both parties agree otherwise, arbitration may be commenced on or after the fifty-sixth day after the day on which a notice of dissatisfaction and intention to commence arbitration was given, even if no attempt at amicable settlement has been made.

Actually, there are some pre-arbitration jobs before an arbitration case. The parties may decide the scope and content of the arbitration, define its procedures and choose the location of the arbitration by specifying these stipulations in the arbitration agreement. Most importantly, the arbitration process bestows the parties the power of choosing the decision maker. That is to say, they can select an individual or several individuals with a specialized expertise in the subject matter of the dispute to listen to the evidence and render a binding decision.

The arbitration agreement either designates the number of arbitrators that will decide the controversy, or refers to institutional rules that provide a procedure to determine the tribunal's composition. In residential construction arbitration, there is usually one arbitrator. In more complex cases, there is usually a tripartite panel, wherein each party appoints one arbitrator and the two party-appointed arbitrators agree on a third. The third is designated as the chairperson of the panel or chief arbitrator because of its neutral position.

An arbitrator must possess the qualification required by the law and by the arbitration agreement. He must:

a. have the legal capacity required by law.

b. possess all the qualifications and none of the disqualifications prescribed by the arbitration agreement.

c. be independent of the parties and thus free from any connection with them and the subject matter of the dispute.

d. be impartial and thus free from bias either in favor of or against any of the parties or any issue in the dispute.

14.1.5 Duty of Arbitrators

The duties of a trained and experienced arbitrator are to:

a. act fairly, independently and impartially, making sure that each party is given sufficient and equal opportunity to present his case, and this, in international arbitration, includes a duty to act in a manner free from national, political and cultural prejudice.

b. weigh the evidence to make a decision. When sufficient evidence is available, the decision maker must weigh the evidence in order to reach a final decision.

c. initiate and plan a management structure in order to resolve the dispute effectively and speedily.

d. proceed diligently taking control of the arbitration proceedings as soon as the appointment has been made.

14.1.6 The Arbitral Proceedings:

a. beginning the arbitration

When parties fail to resolve a dispute, and one party decides to take it to arbitration, the first step is that the parties should read the arbitration agreement and follow the rules agreed upon. Under the ICDR rules, the arbitration is deemed to begin on the date the administrator receives written notice of arbitration from the claimant. The same is true for ICC and LAIA Rules. But this is not always the case. For example, under the UNCTTRCL Rules, arbitration is deemed to commence on the date the request for arbitration is received by the respondent. In some other cases, the arbitration is commenced only after the appointment is accepted by the arbitrators or the request of appointing an arbitrator has been served. The date of beginning is important because it influences how the time of commencement of arbitration is determined.

Under the rules of Arbitration and Mediation Institute of Canada, a party wishing to initiate arbitration shall give written notice to the other party or parties (respondent (s)) of its intention to arbitrate and shall file a copy of such notice with the Institute. The notice shall include:

- the names and addresses of the parties, including the address for service of the applicant;
- a copy of the agreement pursuant to which the arbitration is being initiated;
- a copy of any contract or agreement in connection with which the dispute has arisen;
- a brief description of the claim, the facts supporting it, the relief sought and the amount of the claim.

Institute shall give notice of such filing to the respondent (s) who may file an answering statement in duplicate with the Institute within ten days of the delivery of the notice from the Institute. The respondent (s) shall simultaneously send a copy of the answering statement to the claimant and all other parties. If a counterclaim is asserted it shall contain a statement setting forth the nature of the counterclaim, the amount involved and the remedy requested.

b. preliminary conference

Once the arbitrators have been appointed, the tribunal will usually schedule a preliminary conference or pre-hearing attended by both parties along with their legal council, unless the parties and the arbitrators otherwise agree, to clarify any pre-arbitration issues and discuss the ground rules and procedures to be followed during the arbitration process. If the issues are not considered complex, it may be sufficient to have the meeting by conference call, or by video conferencing. But when it comes to complicated issues which will take rather long time, it might be necessary to have a face-to-face meeting. Considering the importance of this meeting, it is thought to be prudent having experts' attendance, e. g. the lawyers of the parties and an executive-level representative of each party who has authority to make decision for the party.

In the meeting, the parties and the arbitrators will

- determine the issues in dispute.
- determine the matters, if any, on which they are in agreement.
- determine what documents, correspondence, books or records shall be produced, when and by whom, and whether experts are to be called.
- determine the law which will govern the procedures and the substance of the arbitration, unless such law has already been specified in the arbitration agreement.
- consider whether "on site" inspections shall be part of the proceedings.
- decide upon the power of the arbitrator with respect to remedies.
- indicate the number of witnesses likely to be produced.
- estimate the length of time the hearing might take.
- determine whether a stenographic record or other type of recording of the proceedings should be kept or if any particular services, such as interpreters, translations or security measures should be provided.
- determine the manner in which the arbitrator's fee and the expenses of the arbitration will be calculated, secured and paid, including any deposits to be advanced.
- fix the language, date, time and place of the hearing.
- make other determinations that might be necessary before the hearing.
- decide which of the points referred to in the above matters are to be covered by an engagement agreement and complete and sign such agreement either at the meeting or prior to the formal hearing.

At this meeting, the party is allowed to make any necessary motions or objections which can either be ruled upon during the hearing or scheduled for briefing. All pertinent communication should take place at the meeting. After the conference, all communication with the arbitrator must be made jointly by both parties and private conversations between the parties and the arbitrator are generally not allowed and are commonly called "ex parte" conversations which may be a cause of action to vacate the arbitrator's award.

c. written submissions

Prior to the hearing date, all parties shall provide the tribunal and all the other parties with copies of the documents they intend to use at the hearing and the names of all witnesses who may testify on their behalf at the hearing. The initial pleading of the parties could be presented in very detail or succinctly. If the pleadings are succinct, the tribunal may ask for additional written submissions later. On the other hand, if the parties have already submitted extensive information for their pleadings, the tribunal may or may not want additional written submissions to be delivered.

If the submissions essentially constitute the statement of claim and response, they will be submitted sequentially. On the other hand, if the parties are both knowledgeable about their respective positions on the issues as a result of prolific pleadings, but there is a question about which the arbitrator wants to know more, the parties might be required to submit memoranda simultaneously.

The content of the submissions must:

- summarize the claims and defenses;
- identify the significant contested factual and legal issues, citing authority on the questions of law;
- identify proposed witnesses; and
- identify, by name and title or status, the person (s) with decision-making authority, who, in addition to counsel, will attend the arbitration as representative (s) of the party.

d. hearing

Most institution rules either require that the tribunal hold a hearing, or require a hearing if either one of the parties requests it. Parties may choose to give the tribunal power to decide the dispute on documents only. But in practice, most counsels preferred to be heard orally in order to be available to respond to the arbitrator's questions and to satisfy their concerns or provide clarification or explanation. Any dispute not settled at the face-to-face conciliation conference, will move into a more formal phase-the arbitration hearing. Virtually, in all arbitrations, an arbitrator conducts an evidentiary type hearing. Because of the confidentiality of arbitration, the hearing is not open to the public, but only to those who have direct interest in the case. Other persons may attend only by the agreement of the parties and with the permission of the arbitrators. When a necessary party fails to appear in the hearing in person or through counsel, the arbitrator may enter a default award against that party.

The hearing procedures will be:

- Opening statement by chairperson of the Arbitration Panel, to cite authority to hear case and explain reason for hearing.
- The arbitration request will be read into the record.
- The testimony of all parties and witnesses will be sworn or affirmed. All witnesses will be excused from the hearing except while testifying.
- The parties will be given a full and equal opportunity to present evidence and testimony on their behalf and they may call witnesses. The use of expert (e. g. engineer) witness testimony may be especially helpful in cases involving extremely complex or technical matters where the expert witness may be of assistance to the panel in explaining technical practices.

• The parties and their legal counsels will be afforded an opportunity to examine and cross-examine all witnesses and parties.

• The panel members may ask questions at any time during the proceedings.

• The chairperson may exclude any question ruled to be irrelevant or argumentative.

• Each side may make a closing statement. The claimant will make the first closing statement and the respondent will make the final. Prior to closing the hearing the arbitrator shall require each party to state that they have concluded and have no further evidence to offer.

• Adjournment of hearing. Time limits shall begin to run from that point. If written arguments are to be filed, the hearing shall be declared closed as of the final date set by the arbitrator for the receipt of such written arguments at which point the time limits shall begin to run.

• The Hearing Panel will go into executive session to decide the case.

During the hearing, the panel shall have a court reporter or recorder present or may tape record the proceeding. Any party may choose to have the reporter or recorder at its own expense. If transcribed, the transcript should be presented to the Secretary. At the beginning of the hearing, such facts as the time, date and place of the hearing, the presence of the arbitrator, the parties and their representatives, if any, will be recorded.

e. post-hearing proceedings

Normally, the only post-hearing proceedings involve the submission of briefs that may summarize the evidence and arguments. But if one party has turned up new evidence at the end of the trial, sometimes, the other party will have the opportunity to respond to that evidence in a post hearing brief, rather than having to return for another hearing.

If new evidence is discovered by one of the parties after the close of the hearing, but before the award is made, the party can seek to reopen the hearing. The tribunal generally has discretion to reopen the hearings, but is probably more likely, persuaded that the new evidence is relevant and material, to ask the parties to deal with the issues by written submissions.

f. the award

After the conclusion of the arbitration hearings and the submission of any post-hearing documents, the arbitrators' decision or award must be rendered within certain days say 30 days. It may be given orally, but more commonly, it is given in written form and signed by all the arbitrators. No arbitrator shall participate in the award without having attended the hearing. The award must state clearly and concisely the name or names of the prevailing party or parties and the party or parties against which it is rendered, and the precise amount of money, if any, awarded. Awards are usually short, without the need to include a reasoned explanation for how the panel arrived at the conclusion (except when arbitrators, out of discretion, include a statement of reasons for the decision), definite and final as we have discussed in the early part of this chapter. Depending on the parties' pre-arbitration agreement, the award will be either binding or nonbinding. If it is binding, it will be judicially enforceable and reviewable.

If no party has filed a demand for trial *de novo* (or a notice of appeal, which shall be treated as a demand for trial *de novo*) within certain calendar days of notice of the filing of the arbitration award, the clerk shall enter judgment on the arbitration award. A judgment so entered shall be subject to the same provisions of law and shall have the same force and effect as a judgment of the Court in a civil action, except that the judgment shall not be subject to review in any other court by appeal or otherwise.

g. enforcement of the award

Winning an arbitration award may not immediately end the dispute, especially when it comes to the collection of payment from the losing party. It is necessary that the arbitration award is declared enforceable. If the final award can not be timely recognized and enforced by the competent court, the superiority of arbitration in respect of efficiency will certainly be weakened or even thoroughly frustrated.

Theoretically, a properly rendered award by an international arbitral tribunal should be voluntarily respected by the losing party. However, in practice, award debtors, as a losing party, often attempt to resist arbitral awards rendered against them, either because they think the decision to be wrong or improperly rendered, or simply because they believe that they can escape enforcement.

Considering the credibility of international arbitration and the benefit of the prevailing party, the arbitral award must be a meaningful remedy. It is therefore essential that such an award, when properly rendered, be legally binding and enforceable both at the place where it is rendered and at the place where the award debtor has assets.

The New York Convention is today still adhered to by so many countries, one reason of which is that the Convention has been interpreted in a manner favorable to the recognition of arbitration awards. Under the New York Convention, an award issued in a contracting state can generally be freely enforced in any other contracting state. Most of the times, the award rendered should be enforceable, but there are circumstances under which certain, limited defenses are allowed. It happens, for example, when a party to the arbitration agreement is, under the law applicable to him, under some incapacity, when enforcement would be contrary to "public policy" or when the arbitration agreement was not valid under its governing law.

Words, Phrases and Expressions

stipulation *n.* 约定，规定 (regulation, agreement)

bestow *vi.* 给予，授予（grant，give，confer）
diligently *adv.* 刻苦地，勤奋地（industriously，assiduously）
respondent *n.* 被告（the accused party）
counterclaim *n.* 反诉，反索赔（a claim filed in opposition to another claim）
stenographic *adj.* 速记（术）的（of or relating to or employing stenography）
pertinent *adj.* 恰当的，中肯的（appropriate，suitable）
ex parte *adj.* 单方面的（unilateral）
vacate *vi.* 使作废，取消（cancel，abolish）
succinctly *adv.* 简洁地，简明地（concisely，briefly）
sequentially *adv.* 连续的，随……发生的（in sequence，one after the other）
prolific *adj.* 多产的（intellectually productive，fertile）
testimony *n.* 证言，声明，宣言（statement，declaration）
adjournment *n.* 延期，闭会，休会（postponement，suspension of a meeting）
transcribe *vi.* 誊写，抄录，记录（write down）
turn up *vi.* 找到，找出，发现（find，discover）
as of *adv.* 到……为止（since）
de novo *adv.* 重新，开始（from the beginning）
subject matter 目标客体
in duplicate 一式两份
conference call 电话会议
video conferencing 视频会议
engagement agreement 雇佣合同
default awards 缺席裁决
Arbitration and Mediation Institute of Canada 加拿大仲裁和调解协会

疑 难 词 句

1. However，Sub-Clause 20. 5 of FIDIC Conditions of Contract for EPC/Turnkey (Silver Book) regulates that：unless both parties agree otherwise，arbitration may be commenced on or after the fifty-sixth day after the day on which a notice of dissatisfaction and intention to commence arbitration was given，even if no attempt at amicable settlement has been made.

译文：然而，FIDIC《EPC/交钥匙项目合同条件》（银皮书）第 20.5 款规定，除非合同双方另有协议，否则，仲裁将在表示不满的通知发出后第 56 天或此后开始，即使双方未曾作过友好解决的努力。

2. (He must) be impartial and thus free from bias either in favor of or against any of the parties or any issue in the dispute.

译文：他必须公正无私，无论赞成或反对任何一方或任何争端问题都应没有偏见。

3. Institute shall give notice of such filing to the respondent (s) who may file an an-

swering statement in duplicate with the Institute within ten days of the delivery of the notice from the Institute.

译文：该协会应将提交（仲裁）通知交给被告，被告可以于得到通知起 10 日内提交一式两份的答复。

4. The respondent (s) shall simultaneously send a copy of the answering statement to the claimant and all other parties. If a counterclaim is asserted it shall contain a statement setting forth the nature of the counterclaim, the amount involved and the remedy requested.

译文：被告同时应向原告及其他所有参与方发出一份答复。如果被告声明要进行反索赔，他应在声明中包括反索赔的性质、涉及金额及建议的补救措施。

5. determine the manner in which the arbitrator's fee and the expenses of the arbitration will be calculated, secured and paid, including any deposits to be advanced.

译文：确定仲裁员报酬和仲裁费用如何计算、获得和支付，包括提交保证金的形式等。

6. After the conference, all communication with the arbitrator must be made jointly by both parties and private conversations between one party and the arbitrator are generally not allowed and are commonly called "ex parte" conversations which may be a cause of action to vacate the arbitrator' s award.

译文：会后与仲裁员进行的所有沟通都必须要双方共同参加，不允许一方与仲裁员进行私下的会谈，这种会谈通常被称为“单方”会谈，它可能会导致仲裁裁决被废除。

7. (The content of the submissions must) identify, by name and title or status, the person (s) with decision-making authority, who, in addition to counsel, will attend the arbitration as representative (s) of the party.

译文：(在提交的内容当中必须要）提供有决策权的人的姓名、头衔或地位，此人作为该方的代表除了提出建议外，还将参加仲裁。

8. But in practice, most counsels preferred to be heard orally in order to be available to respond to the arbitrator's questions and to satisfy their concerns or provide clarification or explanation.

译文：但是实践中，大多数律师倾向于口头陈词，以便能回答仲裁员的问题，和解决他们关心的问题，或者进行澄清或解释。

9. The use of expert (e. g. engineer) witness testimony may be especially helpful in cases involving extremely complex or technical matters where the expert witness may be of assistance to the panel in explaining technical practices.

译文：如果案件当中包括极其复杂的问题或技术性问题，使用专家证人（如工程师）作证尤其有帮助，专家证人能协助委员会解决技术性问题。

10. If written arguments are to be filed, the hearing shall be declared closed as of the final date set by the arbitrator for the receipt of such written arguments at which point the time limits shall begin to run.

译文：如果要提交书面证据，那么听证会就被宣布关闭以接纳该书面证据，直到仲裁人确定的最后一天为止，从宣布之日起就开始计算时限。

11. The tribunal generally has discretion to reopen the hearings, but is probably more likely, persuaded that the new evidence is relevant and material, to ask the parties to deal with the issues by written submissions.

译文：法庭一般有自由裁量权重开听证会，但如果能使它相信新的证据既相关又重要，它更有可能要求各方以提交书面意见来处理所提出的问题。

12. A judgment so entered shall be subject to the same provisions of law and shall have the same force and effect as a judgment of the Court in a civil action, except that the judgment shall not be subject to review in any other court by appeal or otherwise.

译文：所做出的判决应该受同样的法律约束，应该同民事诉讼的法庭判决具有同样的约束力和效果，只是该判决不能通过上诉或其他方式进行复审。

14.2 Litigation

14.2.1 Litigation or Arbitration

The decision to arbitrate or to go to court is core to the strategy in many contract negotiations and most commercial disputes. In the earlier part of this chapter, we have mentioned many advantages of arbitration against litigation, but one should not take it in the way that arbitration should be adopted in any cases when it is available. For litigation still has its merits, which should not be ignored, in spite of all the disadvantages such as lengthy time, great cost, publicity and so on.

In arbitration, parties should assume that an arbitration award will be final and binding, thus the parties usually have no appeal option, unless an appeal has been included in an arbitration clause. Some arbitration decisions may be reviewed by a judge and may be vacated (removed), if proven to be given on an unbiased basis, which in reality is subject to many limitations. But in litigation, the loser are permitted to express his dissatisfaction and multiple appeals at various levels. Besides, in spite of the experience and competency of the arbitrator, it is still possible that the arbitrator may hold prejudicial opinion in favor of or against one party, which may probably influence the fairness of decision. Comparatively, in litigation, with the judge appointed by the court in stead of chosen freely, the adjudication, to a great extent, will be impartial. Another merit of litigation is that in arbitration, attorneys may represent the parties, but their role is limited; meanwhile, in civil litigation, attorneys spend much time gathering evidence, making motions and presenting their cases.

If the dispute involves technical issues or a small amount, or if the parties want to preserve a commercial relationship, arbitration may be better. If the dispute involves a complex legal issue such as the interpretation of a contract clause or the intent of a law or regulation, or if the relationship between the parties is marked by hostility, litigation may make more sense. And if you strongly believe that the law is on your side, if the stakes are large and if an amicable relationship with your adversary is not a primary consideration, traditional litigation may be a better choice for resolving your dispute. It may also provide an incentive to discuss an early settlement.

14. 2. 2 Process of Litigation

a. The Complaint

In a civil action, litigation begins when the plaintiff files a "pleading" called a "complaint" with the court, which is then "served" on the defendant along with a "summons". The complaint explains what the defendant did (or failed to do) that caused harm to the plaintiff and the basis upon which the defendant should be found legally responsible to the plaintiff. And in the complaint, the demand for the relief sought by the plaintiff should be expressed. The purpose of this complaint stage is simply to determine the legal claims, defenses and other legal issues involved. The pleadings serve as a framework for later proceedings.

b. The Answer

The defendant is given a specific amount of time to file with the court an answer to the complaint. The answer explains the defendant's side of the dispute. Sometimes, the plaintiff responds to the defendant's answer by filing a reply. In some instances, a party may request the other party to clarify or correct its factual allegations or legal theories which are thought to be deficient, and this may lead to amended complaints or amended answers. Once the parties have settled on a complaint, answer, and reply, the case is said to be "at issue", which means that the issues for resolution are now defined.

But if, within the prescribed time frame, the defendant does not file an answer to the complaint, a default judgment is obtained. And once a default judgment is rendered, it is executed.

c. Choice of Alternatives to Litigation

Out-of-court settlement is one alternative to litigation. Indeed, most matters settle before reaching the trial stage. Settlement can be discussed by any party at any time during litigation and there is often a cost-effective alternative to trial.

Two additional alternatives to litigation are arbitration and mediation, which we have introduced earlier. Whether arbitration or mediation is feasible and practical usually depends on the parties' willingness to use these methods. Both the alternatives usually save time and expense, but neither might result in a final resolution of the matter.

d. Discovery

The parties may then conduct discovery, where each side seeks to discover the facts and evidence which are relevant to the legal issues involved and which tend to support or contradict a given party's position. It begins soon after a lawsuit is filed and often does not stop until shortly before trial. in this process, various discovery devices are allowed.

- Interrogatories, which is written questions requesting the defendant to answer under oath within a certain time frame

- Request for production of documents. It is a written request of the defendant to produce for the record copies of pertinent letters, contracts, checks, etc. which the defendant has in his possession and which are being relied upon as a defense to the allegations of the complaint. For instance, if the defendant states they have paid by check, he would be forced to produce the cancelled check.

- Request for admissions. The plaintiff may draft a series of statements, not questions, and send them to the defendant for admission or denial.

- Deposition. It is sworn testimony taken before a shorthand reporter wherein the attorneys can personally ask questions of a party or witness. Depositions sometimes may be used at trial to show inconsistencies in a witness's story or to question the witness's credibility. Depositions sometimes also may be used in place of a witness who is not able to attend trial in person.

e. Pre-Trial Conference

As the trial date approaches, if no settlement has been achieved, courts generally convene pre-trial conference in complex cases such as those involving construction defects. A pre-trial conference is a meeting between the judge and the respective attorneys. The judge will assess the issues of fact or law involved and determine what discovery has been made in order to better plan his docket. He will require the parties to identify all the witness to be called, all documents to be introduced and all uncontested facts that need not to be proven. Quite often, pre-trial conferences are held a few times while the suit is slowly making its way through the court system. In reality, conferences are scheduled by the court in order to hammer out a settlement agreement, thus avoiding trial.

f. Pre-Trial Motions

After the attorneys have thoroughly reviewed the facts, the next step is to file pre-trial motions. A motion is a document asking the Court (i. e. the judge) for a certain action. Motions usually pertain to law or facts in the case, but sometimes they seek clarification or resolution of procedural disputes between or among the parties. In some motions, such as the motion to suppress, the litigant may argue that the evidence the prosecutors want to present at trial does not meet the strict legal requirements and therefore cannot be used against him. Other motions might ask the court to order a party to produce documents or to exclude evidence from trial.

g. Trial

In law, a trial is when parties coming together to a dispute present information (in the form of evidence) in a formal setting, usually a court, before a judge, jury, or other designated finder of fact, in order to achieve a resolution to their dispute.

Normally, a trial includes the following steps: selection of the jury, presentation of evidence, presentation of rebuttal evidence, instruction to the jury, deliberation and decision by the jury, issuance of a judgment.

h. Post-Trial Motions

After trial, attorneys on both parties may file post-trial motions, a motion filed, or deemed filed, after a judgment or a final order, including motions for new trial, reconsideration, re-examination, or modifying or vacating an interlocutory order appealable by right. Usually, the prevailing party will file a motion requesting the court to order the losing party to pay his costs to prosecute or defend the case.

i. Appeal

Following trial, a party dissatisfied with the result may seek an appeal. An appeal is an official request for a higher court to review a trial court decision based on alleged error of procedure or alleged error in application of the law. In civil cases, this usually occurs immediately after a decision is made against one party or at the end of a trial. During an appeal, the parties present in briefs their arguments which then will be submitted to the appellate court along with the record of evidence from the trial court.

Words, Phrases and Expressions

complaint *n.* 控诉，申诉 (accuse; condemning)
summon *n.* 【法】传唤，传票 (subpoena)
allegation *n.* 断言，陈述，辩解 (claim, declaration)
at issue *adj.* 待解决的，争议中的 (under discussion, under debate)
interrogatory *n.* 疑问，讯问，审问 [pl.]【法律】书面质问 (questioning, inquiring)
deposition *n.* 【法】口供，证言 (affidavit, testimony)
convene *vi.* 召集 (organize, summon)
docket *n.* (公文的) 概略，摘要，【法律】判决摘要书，备审案件目录
uncontested *adj.* 无异议的，明白的 (plain, evident)
motion *n.* 【法】申请，请求 (application, request)
pertain *vi.* 关于 (to)，附属于 (to) (be relevant)
litigant *n.* 诉讼当事人 (a party to a lawsuit)
prosecutors *n.* 【法】起诉人，告发人 (chief legal representative of the prosecution)
rebuttal *n.* 【法】反驳，反证 (refutation, confutation)
appealable *adj.* 可上诉的 (applicable for appeal)
alleged *adj.* 有嫌疑的 (not proven yet, suspected)
default judgment 缺席判决 (default award)
cancelled check 已付支票
motion to suppress 证据禁止动议

trial court 初审法院
appellate court 上诉法院

疑 难 词 句

1. The parties may then conduct discovery, where each side seeks to discover the facts and evidence which are relevant to the legal issues involved and which tend to support or contradict a given party's position.

译文：双方接下来可以开展调查，每一方都希望找到与法律问题相关的、能支持或反驳一方观点的事实和证据。

2. After trial, attorneys on both parties may file post-trial motion, a motion filed, or deemed filed, after a judgment or a final order, including motions for new trial, reconsideration, re-examination, or modifying or vacating an interlocutory order appealable by right.

译文：审判结束后，双方律师可以提起审判后请求。这是一种在判决或最后的命令下达之后提起或准备提起的请求，包括请求重新审判，重新考虑，请求修改或废除一项有权对其提出上诉的非正审命令。

3. An appeal is an official request for a higher court to review a trial court decision based on alleged error of procedure or alleged error in application of the law.

译文：上诉是因可疑的司法程序错误或可疑的法律应用错误而正式要求更高一级法院对初审法院判决进行复审。

Appendix: Agreement to Arbitrate

This Agreement to Arbitrate (hereinafter "Agreement") is made this ____ day of June, 2007, by and between STEVEN G. JOHNSON d/b/a ALPINE DISPUTE RESOLUTIONCENTER of P. O. Box 1201, American Fork, Utah 84003 (hereinafter referred to as "Arbitrator"), ________________ of ________________ (hereinafter referred to as "Claimant"), and ________________ of ________________ (hereinafter referred to as "Respondent"). Claimant and Respondent are jointly referred to herein as the "Arbitrating Parties." In consideration of the mutual covenants and promises contained herein and other good and valuable consideration, Arbitrator and the Arbitrating Parties agree as follows:

1. AGREEMENT TO ARBITRATE. The Arbitrating Parties hereby submit the following matter to a binding arbitration with the Arbitrator and have jointly selected the Arbitrator to arbitrate this disputeinvolving ________________________________

____

____

____.

The Arbitrator agrees to arbitrate the Arbitrating Parties' dispute pursuant to the terms of this Agreement and the Utah Uniform Arbitration Act in a fair and efficient manner, and shall serve as sole arbitrator in the case. The Arbitrating Parties hereby waive any rights they may possess to have this matter litigated in a court or jury trial.

2. POWER OF ARBITRATOR. The Arbitrator shall have the power to administer oaths and affirmations to witnesses, to determine the admissibility of evidence 2 and to decide the facts and the law of the case. The Arbitrator shall make all rulings on objections to evidence which arise during the hearing (s) in the matter.

3. ARBITRATOR FEES AND EXPENSES. The Arbitrating Parties shall *each* pay to the arbitrator at least five (5) business days prior to the initial arbitration hearing date the sum of SIX HUNDRED AND NO/100 DOLLARS ($ 600. 00) for anticipated arbitrator's fees and expenses in this case. Additional fees will be assessed if it appears that the hearing (s) will take more time than originally anticipated, which sums are due and payable at least five (5) business days prior to the next arbitration hearing or, if after that date, then such sums are due and payable immediately. The Arbitrator may stay any proceedings until all fee deposits have been made. The Arbitrator may assess fees and costs and expenses to one or both of the Arbitrating Parties as part of the award. If either party fails to pay the Arbitrator's fees or expenses in accordance with the provisions of this Agreement, then the breaching party will be responsible for all costs and attorney's fees incurred by Alpine Dispute Resolution Center in collecting the amount due.

4. APPLICABLE RULES. The parties hereto agree to follow the Alpine Dispute Resolution Center Arbitration Rules as established from time to time by the Center. The Federal and the Utah Rules of Civil Procedure and Rules of Evidence will not apply in this matter, although the Arbitrator may give deference to such rules in order to conduct a fair hearing (s) and to assure that the Arbitrating Parties are allowed due process of law. The arbitrator shall schedule with the Arbitrating Parties any reasonable discovery needed in the matter, and is authorized to limit the types and amounts of discovery as well as the answering times to discovery requests in order to expeditiously resolve the parties' dispute. The Arbitrating Parties shall cooperate fully in providing any appropriate responses to discovery requests as allowed by the Arbitrator.

5. SCHEDULING HEARINGS. The Arbitrating Parties shall cooperate fully in scheduling hearing dates, times and locations, and shall use their best efforts to verify that their necessary witnesses are available to testify at the scheduled times in the hearings in this matter.

6. POTENTIAL CONFLICTS OF INTEREST. The full names and addresses of all parties, attorneys and anticipated witnesses are listed in the attached Schedule A which is made a part hereof by this reference. The Arbitrator is aware of no potential conflicts of interest which may affect his serving as a fair and impartial arbitrator in this

matter, or, in the alternative, the arbitrator has disclosed the following potential conflict to the Arbitrating Parties, and the Arbitrating Parties 3 have determined to nonetheless use the services of the Arbitrator in this matter: ____________________

____.

7. RIGHT TO BE REPRESENTED BY LEGAL COUNSEL. The Arbitrating Parties have the right to be represented by their own respective legal counsel of their own choosing at all times during the arbitration process. Appearance at any hearing without legal counsel constitutes a waiver of the right to have legal counsel present at the hearing.

8. ARBITRATOR DOES NOT REPRESENT EITHER PARTY. The Arbitrating Parties are represented by their own respective legal counsel in this matter, or they have determined to represent themselves in the case without an attorney. The Arbitrating Parties acknowledge and agree that the Arbitrator does not represent either party in this arbitration and that he cannot and will not give legal advice to the parties. The Arbitrator will serve as the judge in this matter.

9. AWARD. The award in this matter shall be a standard award. The Arbitrating Parties agree to abide by and to perform any award entered by the Arbitrator in this matter. Judgment on the award rendered by the Arbitrator may be entered in any court having jurisdiction thereof. [NOTE: in the alternative to a standard award, the parties may elect to have a reasoned award (the reasons for the award are given by the Arbitrator) or an award with findings of fact and conclusions of law. If either of these options is preferred by the parties, the Arbitrator should be notified prior to the hearing (s) in the case.]

10. GOVERNING LAW. This Agreement shall be governed by the laws of the State of Utah.

11. LEGALLY BINDING CONTRACT. THIS IS A LEGALLY BINDING CONTRACT BETWEEN THE ARBITRATOR AND THE ARBITRATING PARTIES. IF YOU DO NOT UNDERSTAND ANY PORTION OF THIS AGREEMENT, YOU SHOULD SEEK AN INTERPRETATION FROM INDEPENDENT LEGAL COUNSEL.

DATED the day and year first stated above.

ARBITRATING PARTIES:

CLAIMANT: RESPONDENT:

____

____

By: ________________ By: ________________ 4

Its: ________________ Its: ________________

ARBITRATOR:

ALPINE DISPUTE RESOLUTION CENTER

By：________________

SCHEDULE A

List the full names and addresses of all parties involved in this arbitration. List the full names，law firm names and addresses of all attorneys involved in this arbitration.

List the full names and addresses of all anticipated witnesses who will likely testify in this arbitration. Attach additional pages if needed.

Glossary

A

abandon *vt.* 放弃(to give up)
abandonment *n.* 放弃(giving up)
abortive *adj.* 失败的
accede *vi.* 同意加入(enter upon)
acceptance *n.* 接受,认可(agreeing to, confirming)
accommodate *vt.* 容纳,接受,允许(to allow, accept)
accommodation *n.* 住处,提供居住的空间或容纳能力(lodgings, space, capacity to receive people)
account for *vt.* 说明,解决,得分(to give reasons for, to resolve)
accountability *n.* 有责任感(being state of responsibility)
acknowledgment *n.* 承认(recognition, confession)
acronym *n.* 首字母缩略词
ad hoc *adj.* 特定,专设,临时(temporary)
adaptable *adj.* 容易适应的,可调整的(adaptive, adjustable)
adequate standing *n.* 良好的名望,声誉,地位(high reputation)
adjoining *adj.* 邻接的,相邻的
Adjournment *n.* 延期,闭会,休会(postponement, suspension of a meeting)
adjudication *n.* 裁决,判决(judge)
administer *vt.* 管理,给予(to manage, to give)
administration *n.* 管理(management)
administration fee 管理费
administrative mechanisms 行政机制
advance payment 预付款(a sum of money the employer pays to the contractor at the outset of a project as a loan for the contractor to mobilize the execution of the works)
advance payment 预付款(down payment)
adversarial *adj.* 对抗性的(conflicting)
adversely adv. 不良地,负面地
advertisement *n.* 广告(ad)
affiliates *n.* 附属机构(Subsidiary bodies)
aforesaid *adj.* 前述的,该(aforementioned)
aggregate *n.* 合计,总计(sum total)
allegation *n.* 断言,陈述,辩解(claim, declaration)
alleged *adj.* 有嫌疑的(not proven yet, suspected)
allocate *vt.* 分配,分派(assign, distribute)
allowance *n.* 补贴,补助(subsidy)
alteration *n.* 更改(change)
Alternative Dispute Resolution 替代性争端解决方法
alternative *n.* 另外的(another)
alternatively *adv.* 作为选择,另外一种选择(otherwise)
amend *vt.* 修订,改正(correct)
American Arbitration Association(AAA)美国仲裁协会
amicably *adv.* 友好地,和蔼地(agreeably, friendly)
amount to 等同于(be equivalent to)
amplify *vt.* 详述(to explain in details)
appeal *n.* 起诉,上诉(taking a case to a higher court for review)
appealable *adj.* 可上诉的(applicable for appeal)
appellate court 上诉法院
apply *vi.* 使用,应用,适用(to fulfil, to serve its purpose, to put into use)
appreciable *adj.* 可观的,大的(substantial)
Appropriate Dispute Resolution 合适的争端解决办法
approve *vt* 核准,批准(to agree officially to)
approved list 经批准(审核)的名单
approximate *adj.* 近似的,大约的
arbitral tribunal 仲裁法庭
arbitrary *adj.* 随意的,独断的,专横的(decided by chance or whim, imperious)
arbitration *n.* 仲裁(arbitrage)
Arbitration and Mediation Institute of Canada 加拿大仲裁和调解协会
Arbitration Institute of the Stockholm Chamber of

Commerce(SCC) 斯德哥尔摩商会仲裁院
arbitration proceedings 仲裁程序
arbitrator *n.* 仲裁人
archaeological *adj.* 考古的
arguable *adj.* 可辩论的(controvertible)
arrive at *vi* 获得(结果);达到(目的)(to reach, come to)
articulate *vt.* 阐明(to clarify, to explain clearly)
artificially *adv.* 人工地,人为地(factitiously)
as of *adv.* 到…为止(since)
as-built' drawings 竣工图
ascertain *vt.* 弄清,搞明白(to get to know clearly, to be aware of)
assembling *adj.* 组合的,装配的(combinational, assorted)
assessment *n.* 评价(evaluation)
assign *vt.* 转让,交给(to transfer)
associated obligation 关联义务
assumption *n.* 假设(to suppose, to accept or believe to be true)
at intervals 不时,相隔一段距离(occasionally)
at issue *adj.* 待解决的;争议中的(under discussion, under debate)
attend *vt.* 照料,照顾(to look after)
attendance *n.* 照管,看管,照顾
attorney *n.* 〔美国〕辩护律师〔lawyer〕
autonomous *adj.* 自治的,自主的(self-directed, self-governing)
availability *n.* 可获得性(being available)
award *n.* 裁定;判决(a grant made by a law court)

B

back-loading *adj.* 向后倾斜的,前轻后重的
balance *n.* 余额 (remaining sum; surplus)
balance sheet 资产负债表
baldly *adv.* 坦率地(frankly)
bar *vt.* 阻挡,阻挠;防止;排斥(get in the way, hinder)
be liable for 对什么负责(be responsible for)
对…应负责任 (be responsible for)
be obligated to 有义务 (be bound to)
be predisposed to 本来爱好…,有…的倾向(be inclined to)
be subject to 以…为条件[转移]的,必须得到…的
bear in mind 记住 (to remember)
bestow *vt.* 给予,授予(grant, give, confer)
betterment *n.* 改善(improvement)
bid bond 投标担保(bid security)
bid price 投标价(offered price)
bill rate 费率
binding *adj.* 有约束力的(involving legal obligation)
borehole *n.* 钻孔 (drill-hole)
breach *vt.* 违反(to disobey)
brief *n.* 诉讼事实摘要,诉讼事件;(律师的)辩护状
buildability *n.* 可建造性 (being possible to be fabricated)
building regulation 房建规则
built-up rate 组合费率(comprehensive rate, all-round rate, overall rate)
bulletin *n.* 公告

C

caisson *n.* 沉箱
caliber *n.* 才干,能力(competence)
cancelled check 已付支票
carry out *vt.* 执行,贯彻,进行,实施(to execute, implement, achieve)
categorically *adv.* 断言的,明确的(explicitly)
caucus *n.* 预备会议;核心小组(ginger group)
ceiling *n.* 上限(upper limit)
certification *n.* 证明,证实(testifying, approving)
China Council for the Promotion of International Trade 中国国际贸易委员会
China International Economic & Trade Arbitration Commission(CIETAC) 中国国际经济贸易仲裁委员会
CIB World Building Congress CIB 国际建筑研讨会 (CIB is the acronym of the abbreviated French(former) name: "Conseil International du Bâtiment"(in English this is: International Council for Building). In the course of 1998, the abbreviation has been kept but the full name changed into: International Council for Research and Innovation in Building and Construction.)
civil action 民事诉讼

civil law 民法
civil Procedure 民事诉讼
claim *vt.* 索赔
claimant *n.* 原告,债权人(plaintiff, accuser)
clarification *n.* 澄清 (making clear or understandable, explaining)
clerk of works 现场监工,工程管理员
code of practice 实施代码,实施规程
code *n.* 法规,章程(statute, constitution)
cognizance *n.* 审理,认定(affirmation)
collaborate *vi.* 合作(to cooperate)
collapse *vi.* 倒塌
combination *n.* 结合(integration)
commence *vt.* 开始 (start, begin)
commencement *n.* 开工(start, beginning)
commission *n.* 手续费,佣金(brokerage)
commitment *n.* 承诺;约定 (an obligation, promise)
common law 习惯法,惯例(practice)
communication *n.* 沟通,交流(exchange)
compensate *vt.* 补偿,赔偿(to pay for)
competent *adj.* 能干的,胜任的(capable, suitable or sufficient for the purpose)
competition *n.* 竞争(rivalship, contest)
competitive tendering 竞争性招标 (selecting a contractor by means of choosing amon many competing bidders)
complaint *n.* 控诉,申诉(accuse; condemning)
completion date 竣工日
comply with 服从,遵从(obey)
compound *n.*(筑有围墙的)院子
comprehensiveness *n.* 广泛,综合(wideness, being embracing)
compromise *vt.* 折中,削弱(to weaken)
conception *n.* 概念,构想(forming an idea)
concession *n.* 特许项目(license, special permission)
conciliation *n.* 调解,抚慰(mediation)
conclusive *adj.* 确定的,决定性的(decisive)
concrete *n.* 混凝土
concurrently *adv.* 同时地(simultaneously)
condition *n.* 条款,条件(term, provision)
conditional payment clauses 有条件付款条款
conditions of contract 合同条款(terms or provisions of a contract)
conduct *n.* 行为 (behavior)
confer *vi.* 商议,协商,谈判(put heads together, negotiate)
conference call 电话会议
confidentiality *n.* 机密性,保密性(the state of being secret)
confidentially *adv.* 秘密地 (in secret, behind the scenes)
congest *vt.* 充满、拥塞(crowed, jam-packed)
consequence *n.* 后果(aftereffect)
consequential *adj.* 作为结果的,由此产生的(resultant)
considerable *adj.* 相当大(多)的(substantial)
consideration *n.* 报酬,补偿费(reward)
constantly *adv.* 经常地,不变地(often, frequently)
constitute *vt.* 构成(to form, to arrive at)
construe *vt.* 解释,理解(understand)
consultant *n.* 咨询工程师,顾问(advisor)
consulting engineer 咨询工程师(the professional consultants advising and supervising the tendering, designing, and execution of a project)
contingency sums 偶发事件,意外事件费
continuing obligation 后续义务(following obligation)
contract administrator 合同管理者,合同管理人员
contract sum 合同额
contractual condition 合同条件(条款)
contractual risk 合同风险(contract risk)
contravened *vt.* 违反 (go against, violate)
contributory negligence 互有过失,原告的疏忽
controversy *n.* 争议 (dispute, argument)
convene *vt.* 召集(organize, summon)
convention *n.*(国际) 公约,协定 (pact, agreement or treaty)
Convention on the Settlement of Investment Disputes between States and Nationals of Other States 关于解决各国和其他国家的国民之间的投资争端的公约
conventional *adj.* 常规的,传统的(traditional)
co-ordination *n* 协调,统筹 (plan as a whole)
corporate *adj.* 公司的,法人的

correspondence *n.* 信函，通信(letter)
correspondently *adv.* 相应地
corrupt *adj.* 腐败的
cost reimbursement *n.* 偿还，退还(repaying expenses)
counsel *n.* 律师，辩护人(lawyer, defender)
counteract *vt.* 对抗，抵消(withstand)
counterclaim *n.* 反诉，反索赔(a claim filed in opposition to another claim)
court action 法律手段
Court of Cassation 最高法院
Court of First Instance 高等法院原讼法庭
cover bid 虚标(a false bid to hide something)
creditor *n.* 债权人
criminal prosecution 刑事检控
criterion *n.* 标准(standard)
crucial *adj.* 决定性的(decisive)

D

damages *n.* 损害赔偿金(compensation)
date for possession 获得工地占有权的日期
daywork *n.* 计日工
de novo *adv.* 重新，更始 (from the beginning)
deal with *vt.* 处理
decisive *adj.* 决定性的
decline *vt.* 谢绝(to refuse)
decree *n.* 法令，政令(statute)
deduct *vt.* 扣除，减去(to take away or subtract)
deductible limit 免赔额 (an insurance term meaning a sum which the insured must pay by himself when settling a claim)
deduction *n.* 扣除，减除(decrease)
deed *n.* 契约(contract)
deem *vt.* 视为(to treat)
default *n.* 违约(breach of contract)
default awards 缺席裁决
default judgment 缺席判决(default award)
defect *n.* 瑕疵，过失(flaw, mistake)
defective *adj.* 有缺陷的
defects liability period 缺陷责任期(maintenance period during which the contractor is liable to make good the defects of the works)
defendant *n.* 被告(the person who is accused)
defendant *n.* 被告(the accused, the respondent)
deficiency *n.* 缺陷，短缺(shortage, insufficiency)
definite *adj.* 明确的，确定的(clear, certain)
degrade *vt.* 使降级
delegation *n.* 委托，授权(entrusting or appointing an authority or power to an organization or individual)
deliberate *adj.* 故意的，蓄意的 (intended)
deliberation *n.* 深思熟虑，慎重 (consideration, thoughtfulness)
demand *vt.* 要求(require)
demolition *n.* 拆毁 (removal)
demonstrable *adj.* 可论证的，可证明的(provable)
demonstrate *vt.* 证明(to prove)
denominator *n.* 共同特性 (common characteristics)
depart from 违背，背离(to disobey)
deposition *n.* 【法】口供，证言 (affidavit, testimony)
deprive *vt.* 剥夺(take away, rob)
derive from 由…起源
description *n.* 描述，形容
designate *vt.* 指定，选定，任命(appoint, select, assign)
designate *vt.* 指明(to point out)
detail design drawings 详细设计图
detect *vt.* 发现，发觉(discover)
deterioration *n.* 恶化；变质；退化 (worsening, decline)
determine *vt.* 终止(to terminate)
dilatory *adj.* 延误的(delayed)
diligence *n.* 勤勉，勤奋
diligently *adv.* 刻苦地，勤奋地(industriously, assiduously)
dimension *n.* 尺寸(size)
dimensioned sketch 尺寸图
disastrous *adj.* 灾难性的(destroying)
disclose *vt.* 透露 (to reveal)
discrepancy *n.* 差异(difference)
discretionary *a.* 任意的，自由决定的 (having one's own decision power, free to act according to one's own judgement)
disputants *n.* 争议者(wrangler)

Dispute Adjudication Board 争端裁定委员会
Dispute Avoidance Panel 争端避免小组
Dispute Board 争端委员会
Dispute Conciliation Panel 争端调解小组
Dispute Mediation Board 争议调解委员会
dispute resolution 争端解决
Dispute Review Board 争议评审委员会
Dispute Settlement Panel 争端解决小组
dispute *n.* 争端,纠纷(dissension)
disruption *n.* 中断(halt)
distinctive *adj.* 有特色的,与众不同的
distinguish *vt.* 区别(to differentiate)
distribute *vt.* 分配(to allocate)
disturbance *n.* 打扰,扰乱(upset)
divergence *n.* 分歧(difference)
diverse *adj.* 不同的(various)
division *n.* 分开,界限,分裂(separation)
docket *n.* (公文的)概略,摘要,【法律】判决摘要书;备审案件目录
domestic subcontractor:DSC 自选分包商
drastic *adj.* 严厉的,激烈的(furious)
due *adj.* 到期的(mature)
duly *adv.* 适当地,适时地 (properly, moderately)
duplication *n.* 重复(repetition)
duration of the contract 工期

E

elect *vt.* 选择做某事,决定(choose to do something, decide)
eliminate *vt.* 消除,排除(get rid of)
elongate *vi. vt.* 拉长,(使)延长(lengthen, make longer)
emerge vi. 出现 (appear)
emergency work 应急工程
emphatically *adv.* 强调地
employment *n.* 雇佣(engagement)
empower *vt.* 授权(to authorize)
endeavour *n.* 努力(effort)
enforceable *adj.* 可强行的(compelled, compulsive)
enforcement *n.* 实施,执行(execution)
engagement agreement 雇佣合同
engagement *n.* 约定,雇佣(employment)
entail *vt.* 使…成为必要,需要(to necessitate)
enter judgment 指做出判决
entitle *vt.* 使有资格,使有权(to give right to)
entitlement *n.* 权利,资格(right, title)
envisage *vt.* 设想(to assume)
equitable *adj.* 公平的,公正的,合理的 (fair, reasonable, impartial)
equivalent *adj.* 相当的,相同的,同等的 (alike, the same)
equivalent *n.* 相等的,相当的(equal)
erection *n.* 建造(construction)
erroneous *adj.* 错误的,不正确的 (incorrect, wrong, invalid)
escalate *vt.* 使逐步上升,逐步扩大 (to grow rapidly, to heighten)
escalation clause 调价条款(fluctuation clause)
essential *adj.* 实质的,本质的(intrinsical, substantial)
establish *vt.* 确定(determine)
established *adj.* 已制定的(set-up)
estimate *vt.* 估算(to evaluate, to appraise)
evade *vt.* 逃避(to avoid)
evaluate *n.* 估价(value)
evaluative mediator 评估式调解人
eventuality *n.* 不测事件(unexpected event)
ex parte *adj.* 单方面的 (unilateral)
exclusion *n.* 排除
exclusive of 不包括(excluding)
exclusive use 专用
execute *vt.* 执行,实施(to perform, implement, or carry out a scheme, a proposal, or a project)
execution *n.* 实行,履行,执行;贯彻(implementation, carrying out)
executive session 执行会议
exercise *n.* 行使,运用
exhaustive *adj.* 全面的,彻底的(all-around, all-sided)
expel *vt.* 赶出;驱逐;开除(drive out)
expertise *n.* 专门知识和技能(professional knowledge and skill)
expiry *n.* 终止;完满;满期 (running out, finishing)
explanatory *adj.* 解释的,说明的(interpretative)

explicitly *adv.* 明白的，明确的(clearly, specifically)
express *adj.* 直接的 (direct)
express provision 明示条款，明文规定(a provision explicitly stated out in a contract, opposite to implied provision)
express term 明文条款(a statement explicitly expressed, opposite to implied term)
expressly *adv.* 清楚地，明白地(clearly, explicitly, opposite to 'impliedly')
extend *vt.* 延长(to postpone, to prolong)
extension of time *n.* 延长工期(extending the time for completion)

F

fabrication *n.* 制作，构成(manufacture, construction)
facilitate *vt.* 使便利，有助于(to help)
facilitated negotiation *n.* 协助谈判
facilitative mediator 促进式调解人
fair valuation 合理的估价
far-reaching *adj.* 深远的
fashion *n.* 方法(way, method)
favour *vt.* 赞同(to agree)
feasibility *n.* 可行性，可能性
feedback *n.* 反馈信息(information, opinion)
final completion 最终竣工
final payment certificate 最终支付证书(the last payment certificate following the settlement between the employer and contractor, which is issued after the engineer receives the final statement and written discharge)
finalize *vt.* 最终定稿，最终敲定
financial guarantee 金融担保(a guarantee by depositing a sum of money)
financial standing 财务状况
finish *n.* 表面光洁度，表面平整度，表面工艺水平
finished item 已完成的项目，成品
flexibility *n.* 灵活性，可变性(changeability)
flow from 起因于，产生于
fluctuation *n.* 波动，起伏(moving up and down)
fluctuations clause 价格波动条款 (escalation clause)
forbear *vi.* 忍耐，克制，避免(refrain, withhold)
forfeit *vt.* 丧失，失去(to lose)
forge *vt.* 锻造，打造(establish)
fraud *n.* 欺骗 (cheat)
fraudulently *adv.* 欺骗地
front-loading *adj.* 向前倾斜的，前重后轻的
function *n.* 职责(duty)
fundamental obligation 基本义务
furnish *vt.* 提供(provide)

G

general(main) contractor: MC 总承包商
goodwill *n.* 商誉；信誉(credibility)
govern *vt.* 管辖，控制(to dominate)
governance *n.* 统治，治理
grant *vt.* 给予 (administer)
gratuity *n.* 赏金，小费(bonus, tips)
ground *n.* 理由(reason)
guarantee *vt.* 保证(to ensure)
guideline *n.* 准则 (rule)

H

hand over *vt.* 移交，交出(to submit)
hasten *vt.* 加速(to quicken)
hearing *n.* 听证会(a chance to be heard explaining one's position)
hereafter *adv.* 今后，从此以后(at some time in the future)
hinder *vt.* 阻碍，妨碍(to impede)
Hong Kong International Arbitration Center(HKIAC) 香港国际仲裁中心.
hybrid *n.* 混合(mix, interfusion)
hypotheses *n.* 假定 (assumption)

I

ICC Commission on Arbitration 国际仲裁委员会
ICSID(International Centre for Settlement of Investment Disputes) 国际投资争端解决中心
identifiable *adj.* 可识别的，可辨认的(recognizable)
immunity *n.* 豁免，免除(being exempt)
impartial *adj.* 公正的(fair)
impartially *adv.* 公正地(fairly)

impasse *n.* 僵局,陌路(deadlock,dead end)
impecunious *adj.* 没有钱的;贫穷的(poor;needy;impoverished)
impede *vt.* 阻碍,妨碍(to hinder)
impediment *n.* 妨碍,阻碍,障碍物(obstacle,obstruction,hindrance)
implication *n.* 含义,暗指
implied term 默认条款
impose *vt.* 强加(to force)
impracticable *adj.* 不可行的(unworkable)
in duplicate 一式两份
in effect 实际上、实质上(actually,virtually)
in general terms 概括地,笼统地(generally)
in respect of 关于,就...而言(on the terms of)
in the capacity of 以…的身份,以…的资格
inadmissible *adj.* 不能允许的,难承认的(unacceptable,not permitted)
incentive *n.* 诱因,动机(motive)
inconsistent *adj.* 不一致的(conflicting)
incorporate *vt.* 使并入
incorporated into 纳入
incur *vt.* 遭受,招致(suffer from)
indefinitely adv. 无定限的,无限期的(for an indefinite period,for ever)
indemnify *vt.* 赔偿,补偿,保护(to compensate for,to make amends for)
indemnity *n.* 赔偿,补偿(compensation)
indices *n.* 指数(复数)
indiscriminate *adj.* 不分好坏的,不加区别的(making no distinctions)
inducement *n.* 引诱
industriously *adv.* 勤劳地,勤奋地(diligently)
inflated *adj.* 高价的,涨价的(high-priced)
inflated tender 高价标(over-estimated tender)
inflation *n.* 通货膨胀(price rise)
infrastructure *n.* 基础设施(public service facilities,such as highway,telecommunication,railway,bridge,ports,airports,etc.)
in-house *adj.* 组织内部的(inner,within,itself)
initial *adj.* 最初的(original)
innocently *adv.* 无辜地
input *n.* 投入(the work,supplies,or effort put in)
insolvent *adj.* 无力偿付债务的,破产的(bankrupt)
inspector *n.* 检查员,检验员
instalment *n.* 分期付款(divided payment)
institute *vt.* 开始(调查),提起(诉讼)(lodge a complaint)
insurance policy 保险单(a written statement of the terms of a contract of insurance)
insuring party 投保方
integrate *vt.* 使…….成为整体,融入
integration *n.* 结合,一体化,整体性(combination,incorporation)
integrity *n.* 正直(honesty,rightness)
interfere with 干预,阻扰
interference *n.* 干涉,干扰
interim *adj.* 暂时的,临时的(provisional,temporary)
interim payment certificate 期中支付证书(the temporary or monthly payment certificate issued by the contract administrator or engineer to the employer for making payment to the contractor for the work done)
interlocutory *adj.* (判决等)在诉讼期间宣告的,非最后的(not final)
International Bank for Reconstruction and Development 国际复兴与开发银行
International Centre for Dispute Resolution(ICDR) 国际争议解决中心
interpretation *n.* 解释(explanation)
interrogatory *n.* 疑问,讯问,审问;[pl.]【法律】书面质问(questioning,inquiring)
interruption *n.* 中断(halt)
intertwine *vt.* 纠缠,紧密联系(connect)
intricate *adj.* 缠结的,错综的,复杂的(complex,complicated)
invalidate *vt.* 使作废,使无效
invaluable *adj.* 非常宝贵的(valuable)
invariably *adv.* 始终不变地,总是
invitation to tender 招标(inviting others to make an offer)
ironmongery *n.* 铁器,金属件(ironware)
itemize *vt.* 逐项列出(to list item by item)
jeopardize *vt.* 危及,损害(harm)

J

job-site *n.* 施工现场(work site,construction site)
joint session【法】联席会议 (joint meeting, joint conference)
joint venture 联营体(a consortium composed of several firms to tender a project as one bidder)
judicial review 司法审查
jurisdiction *n.* 裁判权;司法;司法权(judicial power)
justification *n.* 理由(reason)
justify *vt.* 证明…合理

K

keep to 遵守(to obey)

L

labor relation 劳资关系(relation between employer and employees)
lead time *n.* 完成某项活动所需的时间
legal remedies 法律补救
legal sanction 法律制裁
legal standing *n.* 法律地位(legal status)
legislation *n.* 法律,法规(law, statute)
letter of acceptance 中标函(a letter accepting the tender)
level *n.* 标高,水平高度(elevation)
liberal *adj.* 宽容的,慷慨的(not strict)
licence *n.* 许可证,执照(permit)
lien *n.* 留置权
lightly *adv.* 掉以轻心(carelessly)
likelihood *n.* 可能性(probability)
liquidated damages 误期损害赔偿金 (a penalty which is paid by the contractor to the employer when the time for completion is delayed by the contractor's default)
liquidated damages 误期损害赔偿(a penalty on the part of the contractor who fails to complete a project within the time limit specified by the contract)
liquidation *n.* 清理,清算;(债务的)清偿(insolvency, bankruptcy)
litigant *n.* 诉讼当事人(a party to a lawsuit)
litigation *n.* 诉讼
lodge *vt.* 正式提出(to make a statement officially, to offer formally)
London Court of International Arbitration(LCIA) 伦敦国际仲裁院
lucrative *adj.* 获利多的,赚钱的(profitable)
lump sum 总价包干

M

magnitude *n.* 规模(size)
main or principal contractor 总承包商(a general contractor who signs an agreement with the employer directly and who may sublet parts of the works to a subcontractor)
make good *vt.* 修复(to repair)
mass *n.* 基石(foundation block)
materialize *vt.* 具体化 (to make something general into deails)
measure and value 测量并估价
mechanism *n.* 办法,途径,机制(means, approach)
mediation *n.* 调解(intermediation)
merit *n.* 价值 (worth)
metaphor *n.* 比喻,隐喻(trope)
methodical *adj.* 有条理的
milder *adj.* 温和的(gentle)
mini-trial *n.* 微型审判程序
misrepresentation *n.* 误说,错误的表示或表达
mobilization *n.* 动员
model *vt.* 制作模型,建模(construct a model)
modification *n.* 修改(revision)
monetary *adj.* 金融的,货币的(financial)
motion *n.*【法】申请,请求(application, request)
motion to suppress 证据禁止动议
motivation *n.* 动机,动力,诱因(incentive, inducement)

N

named subcontractor; selected subcontractor 提名分包商
negligence *n.* 疏忽, 粗心大意, 忽视 (inattention)
negligently *adv.* 疏忽地
negotiated procedure 议标程序(a tender procedure by negotiation between the employer and the selective potential contractor)

neutral *adj*. 中立的，不偏不倚的(impartial)
neutrality *n*. 中立(impartiality, refusal to take sides)
nominated subcontractor 指定分包商(a subcontractor who is appointed by the employer directly)
nominated subcontractor:NSC 指定分包商
nomination *n*. 指定，任命(appointment, designation)
non-completion 未竣工
non-litigious resolution 非诉讼争端解决
non-refundable *adj*. 不可归还的(non-returnable)
noticeable *adj*. 明显的(obvious)
notification *n*. 通告书，通知单；报告书(announcement)
notwithstanding *adv*. 虽然，尽管…仍…(although, even though)

O

object *vi*. 反对(oppose)
oblige *vt*. 迫使做
obstruct *vt*. 阻碍(to hinder)
occasion *vt*. 引起(to cause)
odd *adj*. 奇[单]数的 (uneven,(of a number) that can not be exactly divided by 2)
off the shelf 现成的(ready-made)
offer *vt*. 出价，报价(to quote)
omission *n*. 省略，删减(being left out)
open tender 公开招标
operating cost 运营成本，工作成本(working cost)
orderly *adj*. 有序的 (well-arranged)
oriented *adj*. 导向的
origin *n*. 来源，原因 (source)
outsource *vt*. 外包
outstanding *adj*. 未履行的，未偿付的(not yet settled or completed)
over professionalized 过度专业化
overall *adj*. 全面的，总体的(whole)
over-certification *n*. 过度签证(超过实际工程量的签证)
overhead *n*. 管理费用，开销 (expenses, expenditure)
overlap *vt*. 重叠(to lap over)
overlook *vt*. 忽视(to neglect)
override *vt*. 优先于，压倒(to govern)
overturn *vi*. *vt*. 颠覆，毁灭，推翻 (overthrow)

P

package deal 一揽子交易(a set of, a series of, business)
paleontologist *n*. 古生物学者
panel *n*. 小组(team, small group of people)
paramount *adj*. 重要的，首要的(supreme, chief, vital)
partial payment 部分付款
participant *n*. 参与者
particular *n*. 具体情况，详情
pay if paid 只有已经得到付款时才必须支付相应款项
pay when paid 得到付款时支付相应款项
payment Certificate 支付证明
peculiar *adj*. 奇怪的(strange)
perform *vt*. 执行，履行(to fulfil or execute a proposal)
performance security 履约担保 (a guarantee to ensure the contractor well fulfil the contract)
peril *n*. 危险(danger)
persistent *adj*. 持续的，不断的
pertain vi. 关于(to)；附属于(to)(be relevant)
pertinent *adj*. 恰当的，中肯的 (appropriate, suitable)
phase *n*. 阶段(stage, section)
physical *adj*. 物质的
planning legislation 规划法规
pool *vt*. 共享，集中(智慧等)(share, to concentrate idea together)
positive *adj*. 积极的，肯定的
positively *adv*. 肯定地，绝对地(certainly)
possession *n*. 占有，拥有
postponement *n*. 延期，推迟(delay)
potentially *adv*. 潜在地
power *n*. 权力(authority)
practicable *adj*. 可行的，适用的(workable)
practical completion 实际竣工，基本竣工(工程基本竣工后，工程师就可以向承包商颁发接收证书或移交证书)
practice *n*. 惯例 (tradition)

precedent *adj.* 优先的，先决的
preclude *vt.* 阻止，妨碍(to prevent, to disturb)
predictable *adj.* 可预测的(expected)
prejudice *vt.* 不利于，损害(to harm)
preliminary *adj.* 预备的；初步的，初级的 (initial, introductory)
premium *n.* 报酬，奖金，溢价(reward, prize)
pre-qualification *n.* 资格预审(primary selection)
pre-requisite *n.* 前提，先决条件(precondition)
prerogative *adj.* 特权(privilege)
pre-selecting *adj.* 预选的(primary selecting)
presiding judge 审判长；首席法官
presumably *adv.* 大概地，可能地(probably)
presumption *n.* 推测，认定
prevail *vi.* 获胜，有优势(win through, succeed)
prevalent *adj.* 普遍的，流行的(existing commonly, widely or generally)
prime cost 最初成本(initial cost)
proactive *adj.* 主动的(initiative)
procedure *n.* 程序，步骤
proceed *vi.* 进行
proceeding *n.* 诉讼(lawsuit)
proceeds *n.*[pl.] 收入；货款收入，卖得金额；收益 (income, earnings)
procurable *adj.* 可得到的(available)
procurement *n.* 采购，获得(purchase, acquirement)
professional integrity 职业道德
programme *n.* 计划，安排(schedule, scheme)
progress *vi.* 进展，前进(to proceed)
prolific *adj.* 多产的 (intellectually productive, fertile)
prolong *vt.* 延长；拉长，拖长 (extend, put off)
prominent *adj.* 杰出的；显著的；主要的(noticeable, widely known)
promptly *adv.* 立即(immediately)
promulgate *vt.* 颁布(to issue)
pro-rata *adj.* 按比例的
prosecutors *n.* 【法】起诉人，告发人(chief legal representative of the prosecution)
proven *adj.* 经验证或证实的(proved)
provide *vt.* 规定(to stipulate)
provision *n.* 条款，规定(article)
provisional sum 暂定金额有时也叫待定金额或备用金
purport *vt.* 声称(to profess)
pursue *vt.* 追索(to get back, get back)
put forward 提出(to propose)

Q

quantification *n.* 量化(being quantified)
quantify *vt.* 定量
quantity surveyor 物料估算师(cost estimator)
quasi-judicial *adj.* 准司法性的

R

random *adj.* 任意的，随机的
rebuttal *n.* 【法】反驳，反证(refutation, confutation)
receipt *n.* 接受，接收；收条，收据 (reception, proof of payment, note)
receiver *n.* 接收人，接管人，财产管理人
recital *n.* 一系列事件的详述
recognize *vt.* 承认(to accept)
reconciliation *n.* 调停、和解(settlement, compromise, ceasefire)
recourse *n.* 求助对象，采用的办法(resort)
recover *vt.* 收回(to return, to get back)
recovery *n.* 回收，财产收回，追偿(reclaim)
refurbish *vt.* 重建，改建(rebuild)
registered letter *n.* 挂号信
regular *adj.* 连续的(continuous)
regularly *adv.* 定期地，经常地，不间断地(frequently, continuously)
rehabilitate *vt.* 重建，恢复 (to resume, to recover)
reimburse *vt.* 偿还(repay)
reimbursement *n.* 偿还，赔偿，补偿 (compensation, repayment)
reject *vt.* 拒绝 (to turn down, to refuse)
rejection *n.* 拒绝(refusal)
release *vt.* 释放，解除(to free from)
reliance *n.* 依靠，依赖
relieve *vt.* 消除，解除(to release, to remove)
remainder *n.* 剩余部分(rest)
re-measurement *n.* 重新测量
remediable *adj.* 可挽回的，可补救的 (reparable,

retrieval)
remedy *vt.* 修复(to make good, put right)
remuneration *n.* 报酬(reward)
render *vt.* 给予(to administer)
renominate *vt.* 再提名,再任命(to reappoint)
replacement *n.* 替换的人
representation *n.* 表述(expression)
repudiation *n.* 推翻,批判,驳斥 (denial, refutation)
reputation *n.* 名声;名誉,声望,名望 (fame)
reside *vi.* 住,居住(in; at) (be located in, inhabit)
resort *n.* 倚靠,凭借;手段 (option, alternative, choice)
respondent *n.* 被告(the accused party)
responsive *adj.* 符合的,响应的(responding to, consistent)
restitution *n.* 补偿(compensation)
restrain *vt.* 克制,抑制,约束(muffle,restrict)
resultant *adj.* 结果的,合成的(happening as an effect)
resume *vi.* 重新开始,恢复(restart)
retainage *n.* 留置金 (retention payments)
retention fund 保留金(a part of payment withheld by the employer for repairing the defects which might be found later)
retention *n.* 保留,扣留(retaining)
retrospectively *adv.* 回顾地,追溯地(backward)
revert to 归还
review *vt.* 审查,回顾(examine)
revive *vt.* 恢复(to resume)
revoke *vt.* 撤销(to cancel, to withdraw)
rigorous *adj.* 严厉的,严格的(severe)
robotics *n.* 机器人技术
routinely *adj.* 有规律地,通常地(regularly, not unusually)

S

sanction *vt.* 批准,认可(approve)
satisfactory quality 令人满意的质量
scale *n.* 规模(size)
scenario *n.* 情况(setting, situation)
schedule *n.* 资料表
scheduling *n.* 日程安排(setting an order and time for planned events)
scheme design drawing 初步设计图,简图,草图
segregate *vt.* 分离(depart, separate)
seismic load *n.* 地震力
selective *adj.* 选择的(chosen)
semantic *adj.* 语义的 (of or related to meaning language)
sequence *n.* 次序,顺序(order)
sequentially *adv.* 连续的;随…发生的(in sequence, one after the other)
serial contracting 连续承包
set aside *vt.* 留出(to reserve for a special purpose)
set out *vt.* 安排(to arrange)
set something in motion 启动,开始,付诸行动(start, commence)
setting-out *n.* 放线
settlement *n.* 结算,解决
short list 短名单(人数已缩减的候选人名单)
shortage *n.* 短缺(lack)
shortfall *n.* 差额(balance, margin)
shrinkage *n.* 收缩
signify *vt.* 表示…的意思(mean)
simultaneous *adj.* 同时的 (coinstantaneous)
single-point responsibility 单点责任(一方承担全部责任,避免了互相推诿)
skyscraper *n.* 摩天大楼(tower)
sophisticated *adj.* 高级的,尖端的(advanced)
sophistication *n.* 复杂性(complexity)
sort out *vt.* 整理,处理(clear up, deal with)
soundness *n.* 完善,正确(correctness)
specialist subcontractor 专业分包商
specific trade 特定任务
specification *n.* 技术要求,技术规范(technical requirement the workmanship of the works, including the works' materials, permanent plant, parts, and etc.)
spectrum *n.* 一系列,幅度,范围(continuous range)
spur *vt.* 激励(inspire)
standard contracts 标准格式合同(format contract)
standing board 常设争议小组
starting date 开工日(commencement date)
status *n.* 状况,状态(state)
statute *n.* 法令,法规

statutory undertaker 执法人员
statutory *adj.* 法定的(legal)
steadily *adv.* 稳定地(smoothly)
stenographic *adj.* 速记(术)的 (of or relating to or employing stenography)
stigma *n.* 耻辱(shame)
stipulate *vt.* 约定,订定;规定,订明 (demand, specify)
stipulation *n.* 约定,规定(regulation, agreement)
story behind tendering 投标背后的故事,投标潜规则(myth of tendering, unspoken rules)
straightforward *adj.* 简单的,易懂的(simple)
strategy *n.* 策略(policy)
subcontract *n.* 分包合同
subcontracting *n.* 分包.
subcontractors *n.* 分包商(a specialty contractor who subcontracts a part of a project from a main contractor)
subject matter 目标客体
substantial completion 基本竣工(practical completion which enables the employer to issue the taking-over certificate to the contractor)
substantial *adj.* 大量的 (considerable, great deal of)
substantially *adv.* 充分地(fully)
succinctly *adv.* 简洁地,简明地(concisely, briefly)
suitability *n.* 合适,适合 (fit)
summon *n.* 【法】传唤;传票 (subpoena)
supervision *n.* 监督(supervising, overseeing)
survival *n.* 生存
suspend *vt.* 暂停(pause)
suspension *n.* 暂停(stop for the time being)
suspension for cause 有原因的暂停
suspension for convenience 随意暂停
sustain *vt.* 遭受;忍受(suffer, incur)
synonyms *n.* 别名(another name, byname)
systematic *adj.* 系统的

T

take effect 生效 (become effective)
take out *vt.* 获取,办理(obtain)
take over 接收(to receive a project when it is substantially completed)
temporary works 临时设施(site facilities used for execution of a project which are to be removed after the project is completed, such as site office, dormitory, dining hall, batching plant, etc.)
tender document 招标文件(a request for proposal issued to the bidders who prepare and submit a tender accordingly for the sake of being awarded a project)
tenderer *n.* 投标者(a bidder who prepares and submits a tender)
terminate *vt.* 终止(to stop, to end)
termination *n.* 终止,终结;结局,结束(stop, end)
termination for cause 有原因地终止合同
termination for convenience 任意解除合同
termination for default 因违约终止合同
testimony *n.* 证言,声明,宣言(statement, declaration)
the Convention on the Recognition and Enforcement of Foreign Arbitral Awards 联合国承认和执行外国仲裁裁决的公约
The Master Bidding Document for Procurements of Works and User's Guide 世界银行采用的《工程采购和用户指南主导招标文件》
thereof *adv.* 它的;由此 (its)
thereunder *adv.* 在其下;在那一项目[条款]下
tide *n.* 潮水
time-honoured *adj.* 由来已久的
title *n.* 权益,所有权(ownership)
to that effect 有这样/那样的意思或内容
tort *n.* 民事侵权行为
trade contractor 专业承包商,分包商(specialist contractor, sub-contractor)
trade-off *n.* 权衡,平衡,交换(balance, weighing)
transcribe *vt.* 誊写,抄录,记录 (write down)
transfer *vt.* 转移 (to shift)
treasure trove *n.* 无主埋藏物
trial court 初审法院
tribunal *n.* 法庭(court)
trigger *vt.* 引发(start)
tripartite *n.* 三个一组的,三方契约(a team constituted with three persons)
trivial *adj.* 微不足道的,不重要的(insignificant)
turn up *vt.* 找到;找出;发现 (find, discover)

turnover *n.* 营业额

U

unauthorized *adj.* 未被授权的
unbiased *a.* 没有偏见的(impartial)
unconditional *adj.* 无条件的(termless)
uncontested *adj.* 无异议的;明白的 (plain, evident)
under-certification *n.* 不足签证(低于实际工程量的签证)
underestimate *vt.* 估计不足,低估
undertake *vt.* 承担,承揽(to assume)
undue *adj.* 不当的,过分的
unequivocal *adj.* 不含糊的(unambiguous)
unfettered *adj.* 无拘无束的(unconstrained)
unfixed material *adj.* 未固定的材料(在此指'还未被工程使用的'建造材料)
unforeseeable *adj.* 不可预测的(unpredictable)
unilateral *adj.* 单方的
unilaterally *adv.* 单方面的,单边的
United Nations General Assembly 联合国大会
unprofitable *adj.* 无利可图的(unrewarded)
unrealistic *adj.* 不实际的,不实在的(impractical)
unwary *adj.* 不警惕的,易受骗的
utility *n.* 公用事业

V

vacate *vt.* 使作废,取消(cancel, abolish)
vague *adj.* 模糊的,不明确的(ambiguous)
valid *adj.* 有效的(effective)
valuation *n.* 估价(evaluation)
value for money 合算,值得,性价比高(worthwhile, cost-effective)
variant *n.* 变体(derived form)
variation *n.* 变更(a change in the technical specification, work scope or nature of the design, fabrication, and all the works originally specified by a contract)
vary *vt.* 使变化(change)
VAT 增值税
vesting clause 归属条款
vexatiously *adv.* 厌烦地
viable *adj.* 切实可行的,可操作的,可实现的(realistic, workable, practical)
video conferencing 视频会议

W

warranty *n.* 保证 (guarantee)
withdrawal *n.* 撤回,收回(withdrawing)
withhold *vt.* 扣留,保留(retain)
without prejudice(to)【律】不使(合法权利)受到损害;无损于,无害于,不影响
work package *n.* 工作包,工作单元(major piece divided from the work scope of a project)
workmanlike *adj.* 精工细作的(professional)
workmanship *n.* 技艺,工艺(skill shown in a finished craft product, craftsmanship)
works contractor 实际的承包商(working contractor)

References

[1] 周可荣主编. 国际工程管理专业英语阅读选编. 北京:中国建筑工业出版社. 1997.

[2] 张水波主编. 工程管理专业. 北京:中国建筑工业出版社. 2007.

[3] 卢有杰编著. 工程合同管理(双语). 北京:中国建筑工业出版社. 2007.

[4] John Murdoch and Will Hughes, Construction Contracts Law and management, E & Spon, Second Edition, 1997.

[5] The Aqua Group, Contract Administration for the Building Team, Blackwell Science, Eight Edition, 1999.

[6] Michael J. Bond, Esq, *Drafting the Arbitration Clause for International Project Contracts*, Construction Litigation Reporter, April 2007.

[7] Olga K. Byrne, A new code of ethics for commercial arbitrators: The neutrality of party-appointed arbitrators on a tripartite panel. Fordham Urban Law Journal, September 2003.

[8] Philip J. Siegel, To arbitrate or not to arbitrate, Professional Roofing, December 2001.

[9] Ulf Frankel, The Arbitration Institute of the Stockholm Chamber of Commerce, Nordic Journal of Commercial Law, January 2003.

[10] Nael Bunni, The FIDIC Forms of Contract, 3rd Edition, Blackwell Publishing, June 2005.

[11] Margret L. Moses, The Principles and Practice of International Commercial Arbitration, Cambridge University Press; 1st edition, March2008.

[12] Rules of Procedure for Commercial Arbitration, printed by the Arbitration and Mediation Institute of Canada Inc, May 1996.

[13] John W. Cooley , Steven Lube, Arbitration Advocacy(NITA Practical Guide Series)(Paperback), National Institute for Trial Advocacy, 2nd edition, October 2003.

[14] Ross Feinberg & Ron Perl, Construction Defect Litigation: The Community Association's Guide to the Legal Process, Community Associations Press, May 2006.

[15] Techniques for Controlling Time and Costs in Arbitration, Report from the ICC Commission on Arbitration, 2007.

[16] Peter Hibberd, Paul Newman, ADR and Adjudication in Construction Disputes(Hardcover), Wiley-Blackwell, October 1999.

[17] John McGuinness, The Law and Management of Building Subcontracts(Hardcover), Wiley-Blackwell; 2nd edition, March 2007.

[18] Kit Werremeyer, Understanding & Negotiating Construction Contracts(Paperback), RSMeans; 1st edition, September 2006

[19] Arthur Mazirow, Esq. , CRE, The Advantages and Disadvantages of Arbitration as Compared to Litigation, Presented to The Counselors of Real Estate, April 2008.

[20] Christopher R. Seppala, Introduction to the FIDIC Conditions of Subcontract for Work of Civil Engineering Construction,1996.

[21] Sidney Levy, Project Management in Construction(McGraw-Hill Professional Engineering)(Hard-

cover), McGraw-Hill Professional, 5th edition, August 2006.

[22] Keith Collier, Construction Contracts, 3rd edition, Prentice Hall, May 2000.

[23] Talukhaba AA, Mapatha M, Selection framework for domestic subcontractors by contractors in the construction industry, CIB World Building Congress 2007.

[24] M. M. A. Khalfan, P. E. Eriksson, M. Dickinson, P. McDermott, Listing the Voices of Suppliers and Subcontractors! www. irbdirekt. de/daten/iconda/CIB10169. pdf.

[25] The Litigation Process, http://www. tfc-associates. com/news/suit. html

[26] Useful Arbitration Information, http://www. constructiondisputes-cdrs. com/useful _ arbitration _ information. htm.

[27] Introduction to Arbitration, http://www. iccwbo. org/court/arbitration/id4089/index. html.

[28] Commission on Arbitration, http://www. iccwbo. org/policy/arbitration/id2882/index. html.

[29] About ICSID, http://icsid. worldbank. org/ICSID/ICSID/AboutICSID _ Home. jsp.

[30] About HKIAC, http://www. hkiac. org/HKIAC/HKIAC _ English/main. html.

[31] Arbitration award, http://en. wikipedia. org/wiki/Arbitration _ award#Enforcement _ of _ Arbitration _ Awards.

[32] The Litigation Process, http://www. stoel. com/showarticle. aspx? Show=963.

[33] London Court of International Arbitration, http://en. wikipedia. org/wiki.